The Pet Ferret Handbook

D0813895

The Pet Ferret Handbook

Seán Frain

SWAN·HILL
PRESS

First published in the UK in 2002 by Swan Hill Press,
an imprint of Quiller Publishing Ltd.

British Library Cataloguing-in-Publication Data
  A catalogue record for this book
  is available from the British Library.

ISBN 1 904057 06 3

Typeset by Phoenix Typesetting, Burley-in-Wharfedale, West Yorkshire, England.
Printed in England by MPG Books Ltd, Bodmin, Cornwall.

Swan Hill Press

an imprint of Quiller Publishing Ltd.
Wykey House, Wykey, Shrewsbury, SY4 1JA, England
Tel: 01939 261616 Fax: 01939 261606
E-mail: info@quillerbooks.com
Website:www.swanhillbooks.com

Contents

Dedication

To my brother and sister, Michael and Elizabeth, and grand-daughters, Daisy and Yasmine, as wick and mischievous as any ferret!

Acknowledgements

Thanks to John Hill and James Ettles for their assistance with my forays onto the internet and to Mike Frain for some of the photographs. Also thanks to Glynis Frain for her assistance with chasing up addresses included in this book. I also greatly appreciate the confidence put in me by Andrew Johnston. I must also thank Gerry and family, Derek Webster and family, Chris with the kits for their patience while posing for photos.

Introduction

Whichever country you call home, wherever you live – town, country or big city – put simply, ferrets make delightful pets. Their playful, boisterous nature, combined with limitless amounts of charm, will appeal to young or old alike. The way they are often portrayed in the media and the entertainment industry could, however, easily put anybody off wanting to keep them!

The old scruffy rogue standing in the middle of a group of rowdy revellers in some backstreet pub stuffing a ferret down his trousers for a bet, or even a dare, is how these characters are presented to us, with him, often a local poacher who always has the very best 'illegals' for sale, hoping beyond hope that he doesn't get bitten by the 'smelly varmint' whose image is usually of a vicious, wild nature. What utter rubbish. As I have said, ferrets make delightful pets, but a good pet is usually made rather than born and it takes quite a bit of time and effort to end up with a friendly pet, rather than a liability.

Being predators by nature, ferrets certainly have the capacity to become unruly, untrustworthy pets that will bite at the earliest opportunity, but such a creature is often a product of wrong handling, even abuse, or is one that has had very little contact with humans and so has developed a 'wild' nature.

The true nature of the domesticated ferret when it has been well handled with plenty of contact with its human companions, is a real joy to those who seek to keep them. They are good company, and are happy souls who are always ready and willing to play a game or two, and there is absolutely no fear of being bitten. Like a cat or a dog which has been well socialized, it will curl up on its owner's knee and go to sleep. My ferrets have always been well handled and spend much time with the family, thus they are well used to human contact, though if this contact is denied a ferret, then it can indeed bite.

This book will help all who seek the companionship of these charming little animals to ensure that theirs is a pet which enjoys

1

both good health and a good relationship with its owner. This is very important as both owner and pet will be miserable when there are strained relations for whatever reasons.

Ferrets are certainly well suited to the life of a pet animal as they are social creatures which do best in company, so a ferret living on its own, as long as it receives plenty of attention from its owner, is happy enough and will flourish where it has companionship, either human or otherwise. And for those who live in flats or large apartment blocks, the ferret makes an ideal companion.

Dogs are not usually suited to living in such an environment, especially where there are few, if any, places to exercise them, or if they are inclined to be a little yappy and thus disturb the neighbours, causing no end of problems, even threats of eviction. And cats, I think anyway, are happiest where they can enjoy lots of freedom roaming in the big wide world. Ferrets, however, being clean, playful and very social animals, are well suited to life in even the smallest of apartments and are extremely easy to keep occupied and thus happy.

So it is not only their friendly, playful nature which makes them so attractive as people's pets, but it is also that they can be kept almost anywhere, even in a caravan or a trailer home, and will thrive in such conditions, provided they are well cared for, are well handled

A picture of docility; ferrets make great pets.

and have company. This means that, whether you are at school, often with much homework to do almost every night, or you are a part-time or even a full-time worker with an extremely busy schedule, ferrets will make ideal pets.

This book is designed to help you to care for your ferret properly and to make sure that you have a sociable, happy creature which can be trusted with people of all ages. It will advise you of the best methods of keeping these beautiful animals no matter where you live. Far from that smelly varmint which spends its time inside men's trousers or snapping wildly at any fingers that present themselves, you will find that ferrets are wonderful animals which will give you great pleasure and satisfaction for many years to come.

The fact that the popularity of ferrets has risen at an incredible rate throughout America, Canada and much of Europe is, I think, true testimony that ferrets make ideal pets in almost any situation. In fact, ferrets are now the third most popular pet throughout Canada and the USA, and many ferret clubs and societies have been set up as a consequence, with most of them putting on shows. And throughout Britain, country shows are catering for the interest in ferrets by staging both shows and demonstrations. The ferret certainly is a very popular little fellow, and, as we have seen, for very good reason.

Chapter One

History

The domestic ferret has enjoyed a very long association with man. Exactly how long that association is, is impossible to say. I think it is safe to say that man has made use of this extremely versatile creature from the earliest of times – ever since the unending search for much needed food supplies began, when man discovered his hunting instinct, for ferrets were originally domesticated for use in the hunting field, rather than for their charming qualities which make them ideal as pets – a very recent discovery in historical terms.

The rabbit was introduced into Britain during the twelfth century by the Normans as it was familiar to them as a ready source of fresh meat which was essential as they advanced during their conquests. Rabbits were originally 'farmed' in coney beds which were often enclosed. Quite a bit of pasture was always available to them so that they could fatten themselves in preparation of the harvest when these beds (known as 'giants' graves' in England and often marked as such on maps), were finally paid a visit after being left alone in order for the rabbits to breed and produce more meat for the soldiers and the following settlers. I believe that the arrival of the ferret on Britain's shores is inextricably linked with the arrival of rabbits from France, for ferrets were used to drive bunnies from their lairs where they could be caught in nets, or possibly by running dogs, though the former method was the most common.

The ferret had been domesticated by the Romans and would have been used for hunting purposes. That ferret is said to have its roots in the form of the Asiatic polecat, being an albino form of this animal, but whether or not the Romans introduced the ferret into Britain long before the Normans arrived is impossible to say. I think it is more likely that the Normans were responsible for its introduction after the arrival of the rabbit. Later rabbits began to escape from the coney beds, and began to colonize the British countryside where they destroyed farmers' crops and were then hunted to control

A working polecat jill.

their numbers, as well as to feed the people. This massive growth of rabbit numbers in the wild gave rise to a growing popularity of ferrets which found a place in almost every village and town in Britain, being used by poachers and gamekeepers alike. The ferret has an inquisitive nature that makes exploration a must for any self-respecting member of this family, so it has been extensively utilized as a hunting animal for centuries.

Records show that ferrets were in use for rabbit hunting in Britain as early as the twelfth century and this gives credence to my theory that these animals were brought to British shores from France by the Norman invaders. It is unlikely that man had any use for ferrets in the British countryside before the arrival of the rabbit and it is therefore reasonable to assume that it was the Normans rather than the Romans who introduced ferrets to British shores.

Rabbits are prolific breeders and can produce twenty or more offspring during an average breeding season which usually lasts from January until around late August or even early September during warmer summers. Each of these young are themselves capable of producing offspring by the age of about four months, so, as you can imagine, rabbits can colonize a whole country within a relatively short time and they soon became one of the most important food sources the British people had ever relied upon. Rabbits also have excellent fur and this too was utilized for garments, especially during winter. It was for both of these reasons that the ferret was used by man to enter the warrens. The ferrets searched through the dark passages, their inquisitive, nosey nature driving them on to explore every nook and cranny, every inviting tunnel they came across. When they found their quarry they forced them to flee for open country where they would be taken in nets placed carefully over the holes, or by waiting dogs, or, during later centuries, guns, though guns would never be used when the hunter was after the fur as well as the fresh meat.

The poor people always had a need for good, fresh meat and it was the ferret which helped them to get it, though very often the landowner disapproved and many of these poor people were forced to turn poacher, taking both rabbits and gamebirds in order to feed their families and to sell on to make a little money. Every rural community throughout Britain thus had its rogues who tramped the countryside with a rough and ready hunting dog, usually a lurcher (a cross between a greyhound and a farm collie for example), by their side and a ferret encased in a sack. Both dog and ferret would feed on the waste from the rabbit and so would cost nothing to keep.

Many of these poachers were likeable people who were genuinely striving to make a living for their families and turned to their illicit ways out of desperation. Some were tolerated by landowners, as long as they left their more precious pheasants alone and only took enough rabbits for their needs but many poachers, especially those layabouts who didn't want to do a day's work and would rather steal from others for their comforts (nothing has changed), were not tolerated and the gamekeepers would go to infinite pains to catch them and bring them to justice.

While many ferrets escaped into the wild because of badly constructed, or rotting cages, many, if not most, escaped because a gamekeeper would scare off a poacher while he had a ferret inside a warren, abandoning it to its own devices. Of course, these ferrets were well equipped for life in the wild as there was always a very healthy rabbit population for them to live upon, so they soon adapted and many mated with native polecats and these soon colonized Britain, though they were severely hunted down because of their insatiable desire for poultry and young gamebirds. Others were captured and tamed and thus were used once again for rabbit

A black-eyed white belonging to
Derek Webster.

hunting, and it is probably these escapees, albinos said to be
descendants of the Asiatic polecat, mating with native British polecats, that produced what we know as the polecat-ferret, a ferret
which produces both albino and polecat marked offspring in its
litters. These polecat-ferrets are extremely popular and are my
personal favourite, for they are incredibly beautiful and are far
more appealing to my tastes than the albino variety. I once bred an
albino ferret out of two polecat-marked parents, though it may not
have been a true albino. True albinos are all white and have pink
eyes. This one, which cropped up in a litter where all the others
were dark coloured, had pitch-black eyes and was a most attractive
animal.

Though ferrets had proven their great worth as providers of fresh
meat and good fur they now found another use. Rabbits, being the
prolific breeders that they are, soon swelled dramatically in numbers
despite the fact that they were extensively hunted by both man and
beast, and thus soon became a major pest in what is principally an

7

agricultural country, for they ravaged farmers' crops and caused much damage to the rich, fertile pastures which had been carefully tended for fattening up Britain's livestock. Rabbits crop the grass extremely short and thus, in large numbers, will destroy a precious food source for sheep and cattle. Also, their droppings and urine kill the grass until only tough weeds will grow. Added to this is the fact that rabbits love to dig and make many scrapings in the ground, tearing at the roots until grass no longer grows, and you can see why the ferret then became one of the main allies of the pest controller who constantly strived to keep rabbit numbers as low as possible, for rabbits in reasonable numbers are an asset to the countryside, rather than a pest. Where rabbits inhabit ungrazed areas for instance, they keep the vegetation down to a minimum and this aids the growth of low growing and creeping plants which attract some species of butterfly and other insects such as ants. These in turn attract birds and so on, so you can see the important role ferrets have played in managing the countryside.

With the incredibly cruel introduction of myxomatosis in 1954, came a plague which all but wiped out the rabbit from the British countryside and the market in rabbit meat collapsed almost overnight, so many who previous had kept ferrets for rabbitting purposes, now found that it was no longer profitable to keep them and so many, if not most, were released into the wild to fend for themselves and from then until around the late 1970s, ferrets became rather unfashionable until the rabbit population once again began to recover. Numbers once more reached massive proportions in some areas and ferrets again found a place and began to enjoy popularity, but more as a pest controllers' ally, than a supplier of meat, though some still hung onto their ferrets for the purpose of shifting rats from their lairs to waiting dogs who would swiftly deal with them.

It was the brown rat which became a major pest from the eighteenth century onwards, after its arrival on British shores, that was to be the prey of the ferret, second only to the rabbit, so even with the onslaught of myxomatosis which virtually destroyed the rabbit population for many years, some ferrets still found a place with countrymen. Some, as I have said, were indeed released into the wild and this has undoubtedly aided the recovery of the polecat in Britain, for the wild and domestic species have intermingled and it is believed by many that the offspring would eventually revert back to the darker polecat markings. In some places however, such as North and South Uist in the Scottish Hebrides, there are now wild colonies of ferrets which are well established on these islands. They are descended from domestic stock which were kept to help catch rabbits which were an important part of the islanders' diet until myxomatosis reached the islands. The countryman is not inclined to feed an animal which does not earn its keep as he very often cannot afford to do so. Unable or unwilling to keep them, they simply turned their ferrets loose and, being efficient predators, they have survived and become well established. Rabbits are back on these islands in very good numbers and so many ferrets are able to survive because of the abundant food supply. They may not always be very welcome, especially when they get among livestock such as chickens, but, nevertheless, it looks as though they are here to stay.

Just exactly when it was that the potential in ferrets to make good pets was discovered, is impossible to say, though it may have occurred that many who previously kept them for hunting purposes, whether to feed the family, make money, or for pest control, hung onto their animals after the cruel decimation of rabbit numbers, and soon began to see the ferret as more than just a working animal. Children desperately begging their father not to get rid of their ferrets may have been a factor in this and by the 1970s, ferrets were being kept by some as pets rather than as working animals. Ever since then they have grown rapidly in popularity and are now kept much more as pets than as hunting allies, though many still continue to work their ferrets and hunt rats and rabbits with them.

America and Canada have embraced this charming creature during the past couple of decades until it has become one of the most popular pets in both countries. And no wonder, for it has a personality which makes it incredibly suited to the life of a pet, even though it has a very bloody history from the earliest of times.

Ferrets are rising in popularity throughout Europe and I predict that this trend will continue, for they are charming fellows which take little effort and expense to keep. Also, many ferret clubs and societies have arisen and with them shows have grown massively in popularity. These are often staged as money raisers for these ferret clubs, or maybe a ferret rescue centre where unwanted ferrets are cared for, but they are far more than just money raisers. Ferret shows are a great opportunity to meet like-minded people, see other ferrets often in large numbers and, of course, to see how your ferrets

9

This polecat type is very dark and is very similar to a wild polecat, except in temperament of course!

compete with those of other owners. Anyway, more about these shows in another chapter.

The advancement of the domestic ferret has been a long, relentless journey and man has found it a most useful animal. From humble beginnings as an aid to the hunter, the ferret is now, in most cases anyway, highly esteemed as a worthy companion despite the smelly, vicious image that some still try to promote. Whatever the feelings of some people towards ferrets, one thing is certain, they are here to stay!

Chapter Two

The Ferret Family

The Polecat

The polecat (*Mustela putorius*) is native to the British countryside and is famed as a very efficient and ruthless predator. Polecats are also found throughout Europe and have relatives in other parts of the world such as Asia. But it is in the British countryside that it has gained lasting fame – or should that be infamy? – for he is a crafty poacher of poultry and other farmyard fowl, and can cause much damage when he finds himself among them. For this reason polecats were at one time, particularly during the latter half of the nineteenth century, hunted down without mercy in an attempt to curb this severe problem. Also, polecats were removed from shoots which couldn't afford the heavy losses of gamebirds to marauding mustelids.

Polecats were also known by the nickname of 'foumart', or 'foul-marten', because of the glands under the tail which secrete a terrible smelling liquid whenever they feel threatened, thus putting off the attacker from any further pursuit, for that foul odour would tell it that it tasted as bad as it smelt and so killing the owner was point-less. A very hungry fox may have a go at a polecat, or maybe a dog out wandering alone, but few would press home their attack, simply because of that terrible stink – a good defence which was usually very effective.

Because of this love of poultry (polecat translates as hencat through the french word *poule*) and precious gamebirds which were an essential part of the income of large country estates, the wild polecat became increasingly rare until it eventually became extinct in many districts, only clinging onto existence in Wales and a few border counties. The release of domestic ferrets into the wild and the massive decrease in large country residences has meant that the pole-cat is now making a comeback and is found in good numbers in

11

A polecat hob.

many places where it had formerly either disappeared altogether, or had shrunk dramatically in numbers.

Though rabbits play a major part in the diet of wild polecats, the few that remained during the terrible outbreak of myxomatosis in 1954 were not really in any danger of starving to death with the incredibly rapid decrease in the rabbit population, for they have a wide ranging taste and thus many other animals are on the menu. In fact, polecats are not choosy animals at all and even some insects will do. Also on the polecat menu are, for starters, rats, mice, voles and lizards, for main courses, rabbits, hedgehogs and birds, particularly groundnesting birds, and for dessert, eggs, frogs, eels and small fish. As you can see, even with the absence of rabbits, polecats will still eat well and had no problem surviving after 1954. Of course, polecats will undoubtedly thrive better where there are larger numbers of rabbits, but they are not wholly dependent upon them.

Polecats breed just once a year, undoubtedly due to the amount of food available, for the countryside would suffer greatly if it had to put up with double the number of wild polecats, especially when one considers the already high numbers of other predators inhabiting the countryside, and mating takes place usually during April, though many may mate earlier during a particularly mild winter. The young are born around June and consist of five or six young. These tiny little creatures are born blind and with a very thin covering of white hair. They will be marked like the adults by about the age of six or seven weeks.

Though five or six young per litter sounds like a lot of polecats being born in the wild every year, a large percentage of each litter will not survive their first winter. Severe cold, failure to catch enough food, road casualties, becoming victims to dogs not put off by the sickening odour – all are reasons why this is so and it is a sad fact that very few go on to breed the following year. It is no wonder that a relentless campaign to hunt these creatures soon resulted in their near extinction.

The wild polecat is truly a beautiful animal and it is good news that he is making a comeback in the British countryside. Whether any of these wild creatures were ever captured and tamed is difficult to say, but I think it very likely indeed, though it would be incredibly difficult to domesticate them and then use them for rat or rabbit hunting successfully. If captured young enough, then I suppose it may be possible to domesticate a polecat, though I think it would always have a wild nature and thus may not be fully trustworthy. The seeker of the pet ferret, or indeed the hunter of rats and rabbits, would do best by sticking to the domestic ferret for his pleasure.

As its name suggests, the Asiatic polecat (*Putorius eversmanni*) was originally a native of the east but has spread westwards into parts of Europe. This species is said to be the ancestor of the domestic ferret and there is some credence to this theory, for it would certainly have been this breed of polecat that was originally tamed by the Romans, though nations of earlier times may also have utilized the excellent hunting qualities of this eastern polecat. The Asiatic polecat is much lighter than its western cousin and it is believed that albinos thrown up in litters were of the type favoured for domestication and this type remains very popular to this day,

Two varieties of ferret; a polecat and an albino. These are working ferrets.

both for work and as a family pet. The Asiatic polecat has been found in parts of Austria and Poland and is a most attractive animal, though, like his western relative, is not welcome where farming is practised, for the same reasons.

The American Black-footed ferret (*Mustela nigripes*) has suffered greatly from poisons put down for other predators. They have also suffered from prairie dogs, in whose territory this animal is to be found, preying upon them and other small mammals. With black legs and a yellow and brown head and body, it is quite a distinctive animal and, like its cousins, is most attractive and is an efficient predator.

Polecats have managed to spread far and wide and the other varieties are the Russian polecat (*Putorius peversmanni*) and the Marbled polecat (*Vormela peregusna*).

Mink

Though there are quite a few different species of mink, the North American Mink is the most well known. The North American Mink (*Mustela vison*) is possibly best known for its fur which was once in high demand in the world of fashion. Political correctness soon destroyed this popularity, but many people are now tired of being told what to do and fur seems to be coming back in fashion.

Mink have been farmed throughout the world for their fur which at one time fetched incredibly high prices. Mink which have either escaped or been released by misguided and often ignorant animal rights extremists, have successfully colonized Britain's waterways. They are also to be found along Ireland's waterways and in both countries are causing a considerable, even an alarming amount, of damage to local wildlife, for the mink is an extremely efficient, and very ruthless predator which has no natural enemies and has thus spread almost unhindered. Not being native to Britain and Ireland, the American mink is killing off large numbers of aquatic wildlife such as water voles, now very rare in Britain, and coots and moorhen are suffering badly in some places too.

Water voles are very likeable creatures and are extremely graceful swimmers. Kenneth Grahame, author of the internationally renowned *Wind in the Willows*, was obviously charmed by this creature as he modelled his beloved 'Ratty' on it. I first saw one of these beautiful animals while fishing with my dad and brother on a small millpond, and watched, fascinated, as it swam through the water, climbed out onto the bank, its ample body shedding the wet almost immediately, as a duck's feathers do, and disappeared, probably into a hole on the banking. I am deeply saddened that this charming, harmless little fellow is now rare and is in danger of becoming extinct in the not-so-distant future. British waterways would not be the same without the water vole among the wide variety of waterside wildlife. As I write this, in fact, more mink have just been released from a fur farm by animal rights extremists in Hampshire in southern England, in very large numbers, and the devastation that will result does not bear thinking about.

Strict measures of trapping and hunting with minkhounds are in place on some of Britain's waterways though not nearly enough, in an attempt to reduce the mink population, but, it is, I fear, a battle that is slowly being lost and these creatures, beautiful though they are, continue to wreak havoc amongst many species of aquatic wildlife. Of course, in his natural environment in North America, though he still preys upon other creatures, he does nothing like the same amount of damage because he has inhabited his native land for thousands of years and he is a part of the food chain. Not so in Britain where his arrival has been so hard hitting.

Mink produce offspring usually around April or May and have

four to six young, though the numbers can sometimes be as low as two or three, with more being born where there is an abundance of food which consists of fish, birds, rats, mice, voles, frogs, eels and even snakes. Hens and ducks in a farmyard or on an allotment are often a target and in such a situation, like foxes, mink can kill far more than they need for food. They can be found in thick undergrowth usually near to water, in tree roots, drainpipes, in stone piles, or in rat holes which they will take over after killing or evicting the occupants. Apart from their beauty, elegance and their courage, the only favourable thing about mink on alien waterways is that they will help to reduce dramatically the rat population, especially during the summer months when their young need feeding and when families often hunt together in large groups. Rats are even more prolific than rabbits and are disease-ridden creatures, fatal to man and beast, so mink would be more than welcome if it were not for the fact that they reduce the population of birds and small mammals just as dramatically, if not more so.

The Otter

Hunting alongside the mink on British and Irish waterways, but obviously in harmony with the environment as it is a native of these shores, is the elegant and charming creature, the otter (*Lutra lutra*). An incredibly graceful swimmer, the otter, before pesticides leaked into the rivers and reduced his numbers, could be found on every stretch of river, though now, after many rivers have been cleaned up, he is making a very good recovery and is returning to many places where he hasn't been seen for a couple of decades or so. Otters live in holts, holes dug into the riverbank, or under a tangled mass of tree roots, often with the entrance under the water which makes them very difficult to find.

They were hunted with hound and terrier at one time, but hunting them was voluntarily stopped by the masters and huntsmen of these packs because of the falling number of otters and many began hunting mink instead. Otters were hunted because they had a bad reputation for killing large numbers of fish which fishermen feared would decimate the populations of prime fish such as salmon, or trout. Although otters do indeed kill quite large numbers of fish, often wasting some of them by taking a bite out of them and leaving the rest, though other wildlife would obviously benefit from this, I believe that the damage would be very minimal indeed.

Otters are well suited to life in the water as they have webbed feet and a tail which acts as both a paddle and a rudder. Though many books claim that an otter's movements on land are clumsy, he is just as comfortable on land and can get about the countryside very well indeed, often crossing pastures or travelling through woodland on his way between rivers or to a pond or a lake, and sometimes raiding a farmyard in the process, an act which does nothing for his popularity, though these raids are infrequent when compared to those of foxes, mink or badgers. To say he is clumsy on land is utterly ridiculous as anyone who used to follow otterhounds will tell you.

It is during the enchanting month of May that the female otter heads for her chosen holt and gives birth to two or three young which are very carefully and tenderly cared for. The cubs call to one another and their mother by mewing and often have to be dragged into the water, as, unbelieveably, they are naturally shy of water, but, once they have taken the plunge, soon become adept swimmers, expertly negotiating the water and playing games which will help them to catch their prey in the future. One of their favourite games, for either cubs or adults, is to slide down a bank into the water. They will do this over and over again, much the same as a child does (not to mention most adults), at their favourite swimming pool.

Anyone who has read Henry Williamson's classic tale, *Tarka the Otter*, will greatly admire this most charming member of the ferret family and it is a wonderful thing that his numbers are now recovering. This is not in any way due to the fact that otters are now protected and so are no longer hunted, but it is down to the hard work of environment agencies which have made sure that the river systems have been cleaned up. Hunting only slightly dented the otter population as they were very difficult to catch and most simply escaped the attentions of hounds.

The North American otter was hunted for its fur which was known as Virginian. It is found in many places including the north of Canada and Hudson Bay. The sea otter of the north Pacific is much bigger than the common otter and he can often be seen lying on floating seaweed with his paws in the air, enjoying enormously the gentle swell of the ocean and the warm sunshine. Sea otters will often eat their prey while in this position too. Like their European cousins, they are very playful indeed. Otters, whatever type and whichever country they inhabit, are wonderful creatures.

Although the European otter has its cubs usually during the month of May, there is a curious fact which seems to have been overlooked. W.F. Collier, in his book *The Hound and the Horn*, tells fascinating tales of an old Devonshire countryman named Harry Terrell who was much respected for his vast knowledge of wildlife and the countryside in general. Terrell believed that otters bred all the year round as he had seen cubs in every month of the year. Coming from such an authority, one cannot simply ignore this statement which was made well over a century ago.

The Wolverine

The wolverine, or glutton, although found in many countries, is best known in the United States of America, particularly in the north-western parts. It has a reputation as the fiercest of the ferret family and has courage aplenty, attacking many creatures far bigger than itself, in fact, it has been known to attack moose when it has been desperate for food, particularly during periods of heavy snowfall when catching prey becomes difficult indeed. The wolverine (*Gulo gulo*) is very much like a bear in form and is dark brown with a mustard tan on the head and chest. It has incredibly strong limbs which are equipped with equally strong claws, and has a very powerful head with jaws like a steel trap. It is usually hunted for its fur and it has proven a fearless opponent. They will eat lemmings, hares, frogs and fish and will occasionally remove prey trapped in snares by hunters, though sometimes they live on berries much of the time, as the fox will during the early autumn, or fall. It is an animal not to be messed with and is feared even by man, for it could give an excellent account of itself if cornered and threatened.

The Badger

Next comes the badger (*Meles meles*), a charming creature that has reached plague proportions on the British Isles. Badgers live in setts which can be enormous in area, especially when it is an old established dwelling that may have been in use for a hundred years or more, and can be very deep indeed. In fact, in some areas in the north of England they have taken over long-since disused mine-shafts and it is said that some of these extend for miles into the hard rock of the hills. Foxes also share these dwellings and I once watched nine of them emerge from an old mine that was used by badgers.

They are black underneath and grey above with that unmistake-able striped head which means being unable to identify them is an impossibility. They emerge at dusk from their deep abodes and head out to familiar hunting grounds along well worn paths which also lead to latrines – small dug-outs which they use for their droppings. Some say the badger is not at all clean in his habits, but, from my experience of watching these lovely animals on a moonlit night, I would totally disagree with this. True, a badger may put up with a fox in its sett who is a dirty creature indeed, leaving rotting meat and even droppings in and around his earth, but this is usually only a temporary arrangement. The badger will soon evict his lodger when he has had enough of the smell and the mess. He will then set about cleaning his sett by using his powerful digging claws to scrape out the tainted earth, and he will even dispose of his old bedding and replace it with new. Very often strands of grass and dead bracken can be seen leading into the entrance where he has dragged it inside. Make no mistake, badgers are indeed clean animals, as are ferrets, no matter what you may read to the contrary.

Though badgers usually feed on berries, worms, small rodents, young rabbits, whose nests they find with their noses and then dig them up, ground nesting birds, particularly their eggs and young, even honey which they eat regardless of the angry bees whose stings have little effect against the thick fur and skin of the badger, they will also attack young lambs, reared pheasant and very often chickens, for, though they are generally shy creatures, they are also fearless and think nothing of raiding a farmyard. I have a friend who has a farm near Windermere in the English Lake District, and he has lost several of his hens to raiding badgers.

Badgers (or 'brocks' as they are often called) give birth to their young during the early spring and these can often be seen playing around the sett during the summer months, sometimes even during daylight hours. One of the nastier sides to the badger is the fact that they will turn a hedgehog onto its back and will kill it and eat it, leaving just the empty spines as evidence of its crime. Hedgehogs are lovely creatures which control a lot of pests both in the country and in the town where he is a friend of the gardener, so it is sad to come upon the empty spines of a hedgehog which has fallen prey to badgers.

Badgers are now known carriers of tuberculosis (TB) and can spread this disease to cattle which share the same pastures as the foraging badgers. For this reason, badgers are gassed in their setts and this is not a very pleasant death at all, in fact, it is both painful and extremely distressing for badgers which find themselves blocked in their setts while gas is pumped in. Surely government agencies can find a much more humane and less painful way of dealing with TB carrying badgers!

Other types of badger include the ferret badger, the hog badger, and the ratel or honey badger. There is an animal known as a rock badger which is mentioned in the Bible and is listed as unclean food that the Israelites were forbidden to eat. Obviously, the nations round about Palestine used this badger as a source of food which they may have caught using dogs, but it is not a member of the ferret family. It eats vegetation and is about the size of a large rabbit. It lives in holes usually in rock and so is difficult to catch.

Martens

The American marten (*Martes americana*) is very similar to its cousin, the pine marten (*Martes martes*). Though mainly found in Scotland, the pine marten is not native to that country as it once inhabited other parts of Britain and is still found in other countries such as France and Germany. The American marten is differently coloured to the pine marten and is known as sable in the fur trade. Martens have always been hunted for their very desirable fur and so they have been wiped out in many places, sadly, for they are very beautiful and are graceful climbers.

The Weasel

The weasel (*Mustela nivalis*) has a reputation which has suffered greatly during the past few decades, mainly, I believe, because of the way this little creature is portrayed in *The Wind in the Willows*. The book should not be taken seriously as a true study of the nature of the beasts included in the story, for that is all it is, a story, though a very good one at that! Here the weasel is a character with shifty eyes who is always on the lookout for trouble, causing it mainly for others rather than himself, as he attempts to rob Toad of Toad Hall. The shifty-looking criminal is often referred to as a weasel, or a weaselly-looking person definitely not to be trusted. It is a reputation that is totally undeserved and is a slur on this animal's character.

When you consider their small size (between eight and ten inches in length and incredibly slim in body) and some of the prey they will tackle, such as rats, they are very courageous indeed, even putting up a good fight if attacked by creatures such as the fox, or maybe a cat. Generally though, weasels will live on small rodents, hunting them through the narrow passages they inhabit, weaving skilfully below the grass tussocks in search of their prey. Of course, being predators, some of their habits can seem a little unsavoury to the human mind which can be sentimental about animals, and one of their worst habits is to raid nesting boxes, for they can squeeze through incredibly small holes, though, nowadays, it is possible to stop this by putting mesh around the box. They will also prey on birds up to the size of a thrush. During periods of heavy snowfall when prey can become scarce, or maybe during prolonged spells of hard frost, weasels will prey on animals as large as the rabbit and will also fearlessly feed on rats which are bigger than weasels and can bite quite savagely. The weasel is a remarkable little animal which simply abounds with courage and charm. Why he has been labelled with such a bad reputation is really quite baffling.

The weasel will have its young around April or May and then again during late July or early August, providing there is a healthy population of small rodents for them to feed their growing young upon. This second litter is very important as many are killed by motorcars, owls, birds of prey, sometimes by cats and foxes, and gamekeepers too. So their numbers are continually in need of replenishing and the second litter in particular accomplishes this.

Weasels are diligent parents. I once found a nest of young weasels in a pile of rocks on the edge of an old quarry, and I spent an age nearby watching as their mother brought food for them, working tirelessly and nonstop as she returned, time after time, with a tasty morsel for her babies. It is a delight to watch these graceful little creatures as they go about their business in the wild.

Stoats

The stoat (*Mustela erminea*) is a close relative of the weasel and he is a little gladiator, hunting his prey down by scent, prey which is often much larger than itself, and killing it quickly. They mainly prey on rabbits which can kick with very powerful hind legs, potentially killing, or maiming their tormentors, but the stoat will not be put off and will usually overcome its opponent. Foxes will sometimes prey upon stoats during midwinter when hard frost or snow grips the land. It is at this time that stoats can turn white, particularly in Scotland, but sometimes elsewhere too, when their fur is most desirable and is known as ermine, but 'Reynard' will generally avoid this little brawler who is even more courageous than the weasel. They give birth to their young around the same time as the weasel, and, like their cousin, can be seen hunting in family groups throughout the summer months.

I once had the privilege of watching one of these family groups in action. I was out walking with my dogs in a beautiful little valley which nestles at the foot of the broad, sweeping moorland above. Ancient dry-stone walls, walls built without mortar and used for keeping sheep and cattle from getting onto land where they are not supposed to be, climb the hills and give out onto the wild country, following the rough contours of the sides of the valley until they reach the land still not tamed by man. It was along one of these stone walls that I saw this family group out hunting, looking for rabbits which often shelter within the larger recesses of these walls. The youngsters were quite a bit smaller than their mother and all of them chattered warningly whenever I got a little too close to them, otherwise, they were not in any way bothered by my presence and went about their business as though I wasn't there. My dogs were well used to my ferrets about the house and garden, so they took no notice of the stoats as they hunted every nook and cranny of that cold stone wall.

Though you may not be familiar with some of the animals discussed in this chapter, I wanted to include them to give a rounded-out view of the ferret and his relatives (the skunk is also a member of this family and needs no introductions) which are found scattered throughout the earth, highlighting in particular the fine character of these animals, and the fact that they are predators – thus already having a natural tendency to bite. This, I hope, will help readers to see the need to handle their pet ferrets properly; they are well capable of standing up for themselves should they be mishandled or abused in any way. By having a background knowledge of the ferret and the family it belongs to, I am sure you will more fully understand both the characteristics and the temperament of your pet, thus helping you to care for it much more sympathetically.

Chapter Three

Choosing a Ferret

A book on ferrets I was reading many years ago stated that ferrets purchased at any age make good pets, and, while this may be true to some extent, there are some disadvantages to obtaining adults rather than juveniles. Though I have tried hard, I cannot think of one single disadvantage to buying a young ferret, but my own experience is that there are some dangers to starting with adult ferrets. Of course, the more experienced you become at handling ferrets, the more able you will become at handling them, but, for the inexperienced beginner, I would advise, quite strongly, that they begin with a youngster.

It was my brother, Mike, who first began keeping ferrets when we were both young teenage lads. A friend of his had a ferret which was rather wild in nature and had become a little too difficult to handle. Being a teenager with absolutely no experience of keeping ferrets and with no books around to provide guidance, this presented a challenge and my bother took the unwanted ferret off his friend's hands.

It was rumoured that this ferret was actually a wild polecat that had been 'tamed', and I could well believe it, for it was very dark in colour and very wild in nature, but whether or not this was a school-boy tale is hard to say. It was easy to be taken in by this tale anyway, for Sid, (named after Sid Vicious of The Sex Pistols) certainly lived up to his name. It was impossible, or should I say very stupid, to handle him without thick gloves on as he would bite at the first opportunity, and bite savagely at that.

There was one unusual thing that I noticed about him, he was much slimmer and quite a bit smaller than the average hob (male) ferret and I do wonder if this lends a little weight to the rumour that he was a wild polecat, rather than a domestic ferret, for the male in the wild will not usually reach the same size as the tame ferret which, obviously, will feed much more frequently and much more abundantly than the wild animal.

If he was not a wild polecat taken from his natural environment, then I think it safe to say that Sid must have been very badly handled as a youngster, or it may have been that he wasn't handled at all and had very little contact with humans. He would therefore grow up without being socialized and, when he began to be handled, and maybe a little roughly for teenage lads are not the most gentle of creatures, he responded in the most natural way possible, by biting, a ferret's best form of defence. This behaviour would then become a pattern and thus you end up with a totally unsuitable pet, though an experienced handler may well have been capable of stopping him from resorting to biting every time he felt threatened. In the hands of a novice, and a young novice at that, there was absolutely no chance whatsoever of making a success of a bad situation and it was inevitable, if not predictable, that Sid would have to go. He was passed on to another of my brother's friends and I do not know what happened to him after that.

This episode during my early years taught me important lessons: the necessity of being informed about the pets we keep, hence the need for books of this kind, and that adult ferrets, unless they are well socialized and are used to being handled regularly, do not make suitable pets for children. It is always better to allow youngsters to begin their ferret keeping with ferrets fresh from the nest, unless, as I have said, the adult ferret is very familiar with human contact and has proven itself a non-biter, which, it has to be stated, is what most ferrets are. Just as good pet ferrets are made rather than born, so too is the biting ferret made rather than born; abuse usually being the chief cause of a ferret's tendency to bite.

This was to be the one and only ferret that my brother kept, but this experience planted a seed in me and it wasn't long before I gave up keeping and breeding mice as pets, and moved onto ferrets instead, despite my fairly brief association with Sid Vicious and his attempts to bite which always found him hanging on grimly as he tried in vain to bite through the thick glove.

I was just thirteen years old when I saw the two polecat-ferrets in my friend's back yard, two small jills (females) which were extremely attractive animals and were full of irresistible charm, a quality ferrets have in abundance, and I knew then that I just had to have one of them, especially when I saw how easily handled they were in comparison with Sid. From then on I pestered my friend until he at last agreed to sell one of them to me. So I at last purchased my first ferret, but this proved to be a complete disaster, the tale being told in full in the chapter concerning the housing of ferrets.

Pestering my friend again, I eventually was able to purchase the other ferret which I named Judy. She was one of the darkest coloured ferrets I have ever seen, after Sid that is, but at least she was friendly, although after my experiences with Sid I preferred to continue wearing thick gloves whenever I handled her, just in case!

It was a continuous learning curve with this ferret, there being a definite shortage of books on the subject at the time (even now, though there are many books which cover the working of ferrets, there is still a shortage of good books covering the keeping of ferrets as pets), but it wasn't long before I had purchased another ferret, Ben, a huge hob who looked more like a bear and who had a massive head, the biggest head I have ever seen on any ferret. He was a real character and loved to play and by this time I had grown in confidence and had shed the gloves, and watched him bounding sideways, his huge frame and massive head bobbing up and down as he played his silly games with his female friend, or with myself or my brother. I learnt a lot from these two and it wasn't long before I was handling them well and began to trust them much more. I really enjoyed keeping and looking after them, so much so, in fact, that I decided to invest in another ferret, an albino, which I bought from another friend.

I had found Ben and Judy to be very friendly, playful animals which had absolutely no malice in them at all. In contrast, this all white ferret whose name, for the life of me, I cannot remember, had no inclination to play and seemed to be quite a solitary animal, keeping herself to herself, fixing me with a cold eye which should have rung warning bells, especially after Sid who could fix you with the coldest of eyes, but I persevered with her and hoped she would respond to my patient handling. It wasn't that she was a biter, not at all, she had shown no tendency to bite at all, it was just that she was so unfriendly and unsociable.

I used to exercise my ferrets in my bedroom a lot of the time, a very large room which I shared with my brother, and one day Judy and this white ferret were out and about, running all over the room, climbing onto the beds and then jumping off again, exploring the dark passages behind the drawers and the wardrobes, or slithering along the narrow tunnels I had made for them using my large collection of books which I piled up on the bedroom floor. They loved to

25

wander through these tunnels and I thoroughly enjoyed watching them enter the darkness of these winding passages and then attempting to guess correctly which exit they would emerge from.

My brother was watching them with me and he loved to tease them. While he was playing with Judy, the white ferret climbed up onto the bed next to him and walked, rather unconcerned, to the edge. Mike was bending down and doing his usual teasing routine with Judy, totally unaware of the other one who watched him with a keen interest which was unusual for her, for she never really took much notice of anything. There must have been something in my brother's poise which caught her attention, for she suddenly leapt from the edge of the bed, flying gracefully through the air at an amazing speed and for some considerable distance until she landed with her teeth fastened firmly through my brother's ear lobe. What on earth motivated her to do such a thing I do not know, but, if my brother had chosen to do so, he could have put two earrings in his left ear lobe after that incident, in fact, just for a brief few seconds, it looked as though he had one of those long dangly earrings in his ear. The only difference was that this one was alive and had no intention of becoming a fashion accessory!

Thus the need for caution when choosing a ferret was well illustrated in both of these accounts concerning the choosing of adult, rather than young, ferrets when starting out in keeping them. It would have been far better for me had I begun with a young ferret, known as a kit, until I had become more adept at handling them. So be careful about advice given which says that ferrets of any age make good pets. Unless you know the history of a particular animal, who has had it, why its owner is parting with it, how well handled and socialized it is, etc, etc, then my advice is always to go for a kit, especially if children will be involved in the husbandry of the animal. However, in order to handle a ferret, you must first know from where to obtain one.

Pet shops, particularly in Canada and America, are good places from which to obtain your ferrets, some even specializing in selling descented ferrets (ferrets with their scent glands surgically removed so that they do not smell as pungent) or ferrets which are of a strain noted for their friendly nature, making them much easier to handle. In Britain, except in those pet shops which sell more exotic pets, ferrets are not as readily available but there are plenty of other sources from which ferrets can be obtained.

You may wish to buy from a breeder and scanning the livestock ads in your local papers may pay off, although you are more likely to find the ideal pet fitch (another name for a ferret) if you consult those publications which are likely to contain advertisements of this kind. For instance in Britain publications such as *Loot*, *The Countryman's Weekly*, *Exchange and Mart*, among others, often carry adverts for ferrets, though many of these young are bred from working parents, that is, from parents which are used for the hunting of rats, rabbits, or both. Does it matter that they are bred from hunting animals? Not at all. Whether ferrets are bred from pets or workers has absolutely no bearing whatsoever on the temperament of the youngsters, for workers need to be friendly and easily handled if they are to be of any use to the ferreting man or woman. All of the people I have known who hunted with their ferrets had easily handled, well socialized fitches which made ideal pets, as well as working animals. So do not be put off if you have the chance to buy a youngster from working parents.

In the Appendices I have included a list of addresses of clubs and societies in as many countries as possible and contacting one of these nearest to you may help you to find your ideal pet if you are having difficulties finding a suitable breeder. Also included are the addresses of welfare societies which often run a ferret rescue centre, that is, looking after previously unwanted ferrets which have either been abandoned, handed in, or, in some cases, released into the wild to fend for themselves and have been recaptured again after turning up at someone's house. If you are intent on purchasing an adult ferret, then you cannot go far wrong if you contact one of these centres and go along to choose one of their animals. You can be assured that these ferrets are well cared for and are well socialized and thus will make ideal pets. Even if you are searching for a child's pet, adult ferrets from a rescue centre may be ideal, for the person in charge of the rescued animals will know them well, especially if some of the animals have been with them for any length of time, which some will most certainly have been, and will be able to recommend those fitches which are suitable to be handled by children. It may be that some of these centres have children helping out and so their animals will be very familiar with kids, and thus will be well used to being handled by them. In fact, for you young readers in particular, if there is one of these centres nearby, then why not volunteer to help out? This would be an excellent way of getting to know more about ferrets and their ways, especially their care, and this in turn will better equip you to handle and care for your own fitch.

Pet shops, local newspapers, publications which are noted for advertising them, and clubs and welfare societies, are all excellent sources from which to obtain your future pet. With sources such as these, I think it quite impossible not to find the ideal pet for you. Now that you have some idea as to the age of your intended pet, either a kit or an adult, and from where you can obtain it, it is time to decide on the sex and the colour of the animal.

Male or Female?

The hob ferret grows much bigger than the average female and, by the time he reaches adulthood, will be around fifteen or sixteen inches in length, not including his tail, and, it has to be said, will be a lot smellier by the time he is full grown, on account of the need of the male to mark his territory in the wild. This would serve two purposes, firstly, it would keep all but the most belligerent of males off his hunting grounds, thus preserving his food supply to maximum effect, and, secondly, to try to keep other males from his mate during the breeding season. So this smell which a lot of people find very offensive (it is a very pungent odour) is only natural and is part of the make-up of both male and female ferrets, although females are far less smelly than their male counterparts. This odour does tend to cling to clothes and even skin to some extent, so, if you feel that you cannot put up with this natural ferret odour, then you would be better off having your ferret descented, that is, having its anal glands removed by your local veterinary surgeon, or by a vet who specializes in this procedure on account of their being plenty of ferret owners within the area he covers. Remember however that females do not smell all that badly and I would not recommend this operation in their case, and I would only recommend this for males which are to be kept permanently in the home, rather than out in the garden. Descenting may also be best if you choose a male who, although living outside, spends a lot of time inside your home with you, as many do. It really is a powerful odour and it clings, so a home occupied full time, or on a regular basis, will soon begin to smell and you may find that visitors come far less frequently, though this may be an advantage in some cases! For a hob to be successfully descented, he must also be neutered, as removing the scent glands alone is not nearly effective enough, unlike the female, for she will only need the scent glands taken away. If you wish to breed from your hob in the future, then you will either have to put up with the smell if you have him indoors, or you will have to keep him outside if at all possible, for the latter option is by far the best. The reason for the male ferret needing to be castrated also is because his scent is also for attracting females during the mating season, as well as for territorial reasons. Polecats in the wild can range far and wide and this musky odour will help him both to find a mate, and to woo her, thus the need for castration too. Though this procedure may be necessary to ensure a pleasant smelling home, I would never recommend castration for a ferret with behavioural problems, a biter for instance, as the chances of it making any difference whatsoever are very minimal indeed. In fact, I have yet to see such a procedure have any benefits on an animal with bad habits, horses excepted, and a dog I once had which was totally untrainable and was very aggressive, retaliating when disciplined, went even worse after he had been castrated.

Why the male is far bigger than the female (jill ferrets reach between twelve and fourteen inches in length) is a mystery, but this large size of the male has been well used and taken advantage of by rabbit hunters for centuries. The hob ferret used to be known as a liner, or a line ferret due to the fact that he was attached to a line while he was put down a rabbit hole. This wasn't to bolt rabbits into

A hob (left) and a jill ferret. The difference in head size can clearly be seen.

nets, for many feared that few rabbits could bolt from the larger male whose bulky frame would almost certainly fill the dark passages and give little chance of a rabbit pushing past him. The jill ferret would usually be used to hunt and bolt the rabbits from below ground, but, occasionally, she would kill her quarry inside the warren. Some would try muzzling their jills in order to stop this from happening, but this wasn't really an option for those who had their ferrets' best interests at heart, for they could meet a rat in the darkness and, obviously, would not be able to bite it. The ferret, especially if the rat was a female, guarding her young ones, may then be seriously injured, unable to do a thing about it. Also, a muzzled ferret would not be in any position to stop a rabbit from kicking it with its powerful back legs, as the kick of a rabbit is potentially lethal to a small jill, especially if the kick hits the right spot, so jills were and are generally worked without being muzzled, thus they do sometimes kill, rather than bolt, their quarry.

The large hob would then have a collar fixed around his neck, as tight as possible to reduce the risk of it snagging, the collar had the line attached to it, and he was then released into the warren where he would search for the kill, drive the smaller female away from it, who would then emerge where her owner was ready for her, and the male would then be pulled out, sometimes dragging the rabbit, at other times without it. Of course, nowadays, ferret locators have made the line ferret redundant as the collar bleeps in the receiver and the rabbit hunter digs down to recover both his ferret and her prey. Hobs are still used for hunting rabbits however, though far less frequently than jills.

When young children are involved in both the husbandry and the handling of ferrets, the smaller jills may be more suitable, as the hob is also much heavier and thus is more difficult for younger kids to deal with, though both sexes are just as gentle when they have been well socialized as babies.

In some cases, the male may even be gentler than his female counterpart, and Ben, the ferret mentioned earlier, was incredibly gentle and had not a bit of malice in him. In fact, I have always found that the more spiteful, short-tempered ferrets have been females. I have never had a problem from any of my male ferrets, all have been friendly and playful and full of fun, and have been a pleasure to keep. Having said that, it is only a small minority of females which have caused any problems. Though I have had scores of jill ferrets, very

Charlie Dunkon
with her albino
ferret, 'Bambi'.

few, only one or two in fact, have ever bitten, or threatened to bite, so maybe I am being a little unfair.

Ferrets, both males and females, have varying personalities and some can be downright moody. While one may be very friendly, another may be very standoffish, even totally unsociable. I once bought a polecat ferret of the variety known as a pencil ferret, an incredibly slimline type which favours a weasel in body shape, and I named her Wiswell, which is the old English word for a weasel, and she was very unfriendly indeed. She never bit me, or even tried to bite, it was just that she wasn't impressed at all with human company and would go about her business doing her best to ignore me. One of her sons inherited this same trait and he too was very slim and was quite a bit smaller than the average male. In fact, they were so slim it looked as though they were under-nourished. No matter how much I fed them, they refused to put on any weight and

31

I have to say, due to their unsociable nature, I found it difficult to take to them, no matter how hard I tried.

Others, though you couldn't label them as unsociable, are rather indifferent towards their owners and can take them or leave them. Some will go to sleep in your lap, while others wouldn't be seen dead cuddling up to their owners. Like us humans, ferrets' personalities differ greatly and it is all part of the rich tapestry of life.

Choice of Colour

Ferrets also come in a variety of different colours including chocolate and red and you may have a preference for a certain one, such as the albino. My preference is for the darker polecat colouring which, in some cases, is nearly as dark as the wild variety. I think that this is a most attractive colour. The top coat is almost black with an off-white undercoat which shows through just slightly and this gives them a grizzled appearance. They have white tipped ears and a dark bar across the face, taking in the eyes, which are dark and very attractive, with white patches on both sides of the head and white around the nose. Their tail is usually as dark, if not darker, as the body fur and, when brushed out, makes the appearance even more appealing. In some cases, it is difficult to tell a wild polecat from a domestic one and, though nowadays all polecat marked ferrets are known as polecat-ferrets, when I began keeping them during the late nineteen seventies, they were commonly known as polecats in my area, for there is very little difference. The dark variety are usually known as sable, but I have never got out of the habit of calling them polecats. I knew many rabbit hunters when I began keeping ferrets and most, if not all of them, believed that this darker coloured fitch was undoubtedly the best for bolting rabbits.

The lighter coloured polecat-ferret has become known as a Siamese or a sandy ferret, but I have always known this as simply a polecat-ferret, because it is marked as though it is a cross between an albino and a polecat. This variety has a lighter coat, still with some darkness to it, but the lighter undercoat shows through much more. The face is also much lighter and the dark bar much less prominent when compared to the sable, or the dark polecat type. Their legs are still usually very dark though, and I think this is also very attractive. I once had a beautiful ferret of the Siamese variety named Numbhead (she was beautiful, but rather dim and slow on

A silver and a
polecat type.

the uptake, hence the name) who was killed by a friend's terrier during a regrettable lapse in concentration. She had a wonderful temperament and was one of my favourites. I was devastated.

The next variety is the albino ferret, a very common variety simply known as an albino. I have never really taken to this type, but, nevertheless, they are also very attractive and most people find them appealing. Sometimes you get an all-white ferret with black eyes which look as black as black can be surrounded by a sea of white fur and this type is irresistible. Whenever I have bred such a ferret, I have just had to keep it, for they are not only very beautiful animals, but are most unusual. The albino, when I first started in ferret keeping, was known commonly as a ferret, as opposed to an albino. I may have just been unlucky, but I have found this variety to be less friendly than the darker varieties, but then I have had far more of the dark coloured ferrets than the albino variety, so I may not be a good judge on the matter. Had I had the same number of whites as I have the darker ones, then maybe I would come to a different conclusion.

What I know as the silver-mitt ferret is commonly known as the white-footed ferret in America. They have white feet and a white patch on their neck. Sometimes they have white on their legs too. Though some of these white-footed ferrets are dark in colour, sometimes just as dark as the sable, or polecat type, most are usually silver in colour and, I have to say, in my opinion, are not as appealing as the lighter silvers which are now very popular. You may favour them, but the only silver-mitt ferret I owned, though of an excellent disposition, wasn't very attractive in appearance. This colour is due to white mixing in with the darker hairs and giving a silvery appearance, though some would say that it was simply grey in colour.

I saw my first silver ferret at a local auction many years ago. She was a small kit, maybe of around eight weeks of age, and I couldn't resist buying her, even though I had plenty of ferrets at home and wasn't really looking for another, but her markings were most unusual and I had never seen one like her before, for silver ferrets were rather rare in my neck of the woods at that time.

The Littleboro' auctions, situated below the wild, towering moorlands near Rochdale in Lancashire, England, was a place where you could buy anything – dogs, antiques, furniture, all kinds of exotic animals – but it was an especially good place for finding a wide variety of ferret and usually at very reasonable prices. I named my silver ferret Titch, for she was a tiny thing, even for a kit, and, even in adulthood, was much smaller than the average fitch.

Apart from the most common type already mentioned, there are the pencil ferrets which are built like weasels, as was Wiswell, and some of these can be extremely small in size when full grown, though they are far from common and may be very difficult to obtain. I don't really think the size matters. As long as a ferret is of the correct temperament, then that is the most important thing.

Choice of Ferret

Having found our source from where ferrets can be obtained, having decided on the sex of our pet and having settled on a colour which appeals the most to us, it is now time to go out and pick one. Of course, though you may prefer a certain colour, there may be a ferret which appeals to you more that is of a colour you least favoured, so go out to choose your pet, yes, with certain ideas, but with an open mind.

When you are making your choice, whether it be an adult or a kit, you should look for good health as a starting point. He or she should be very alert and active, full of curiosity or just downright noscy, approaching you without hesitance and with a friendly nature.

Avoid the shy, retiring type, for these can turn out to be biters, for they are nervous creatures and soon feel threatened, the reason why many ferrets bite. True, with plenty of patience and careful handling, the ferret, adult or kit, may overcome its initial fear and trepidation, but that one is best left alone, unless you have plenty of experience with fitches. If I was a breeder who was selling such an animal, I would give it extra attention and handling and, especially if it was a youngster, I would keep it on a little longer until it grew in confidence. The breeder of ferrets should know better than to try to pass on stock with undesirable traits.

The one which is the boldest and friendliest is a good option, although, of course, that particular one may not be of the right sex or colour. You just have to follow your instincts. Make sure the eyes are bright and shiny, clear of any matter and full of inquisitiveness. The fur should be clean and on the bushy side. It should not be dull and lifeless, thin and straggly, or flat to the body constantly. A good way of testing that a ferret is well nourished is to gently pinch the skin at the back of the neck (a place a kit should be well used to being handled as this is the area where the mother will pick up her youngsters) between finger and thumb and gently pull it. The skin

A beautiful polecat type.

35

should immediately go back to its normal place. If it remains stuck out, then there is something wrong. Also, a ferret should carry a bit of surplus weight. If a kit is skin and bone, then avoid buying, for it may be what is termed in livestock circles, a 'poor doer', or it may simply be that the breeder, or maybe even the pet shop owner, is not caring for them properly. It could also be that an undernourished kit is the runt of the litter and is being beaten to its fair share of the food by its other, much stronger litter mates (see the chapter on Breeding for advice on caring for the runt of the litter).

To summarize, a young or an adult ferret should be bright-eyed, full of energy, curiosity and fun, should have clean, bushy fur and carry a good amount of weight, much more weight in fact, than their wild relatives will carry, for fat is quickly turned into energy as the wild polecat, or ferret, goes on its long and endless search for food. If you find all of these qualities in a litter or in an adult ferret you have gone to see, and you have picked the one which most appeals to you, maybe a white male when you were thinking of a sable female, then why not go ahead and buy?

When a breeder has a good reputation for supplying well reared and well cared for, friendly fitches, it is fitting to purchase a kit, for he or she will be jealous of their reputation and will do their utmost to maintain it, carefully guarding against passing on any poor doers, or any youngsters which are inclined to be unfriendly, even aggressive, when in contact with humans, and, more importantly, children. As we have seen, ferrets can give a very nasty bite, even through to the bone when they mean business, though this is very rare. Of course, some kits are inclined to nip a little, but, with plenty of contact with its owner, will soon grow out of this, for it is simply following its wild instincts, for young polecats will taste many things as they learn about suitable and, indeed, unsuitable foods, but it is something the domestic ferret will quickly get over. So do not think that a kit nipping, or mouthing your fingers is a sign of aggression, for it is simply exploring and learning about life.

As yet, you may be undecided about whether or not you will breed from your ferret. Descenting the male by removing both the anal glands, or scent glands, and by castrating him too, can virtually be done at any age (although some vets advise waiting until the hob is sexually mature before castrating), as can the spaying of the female. If you cannot make up your mind do not despair, for you have time to think about it and, especially where kits are concerned, the smell does not reach its full pungency until they reach adulthood and sexual maturity, so, even in the case of a hob, there is no rush. If you have purchased an adult, then you will already be familiar with its odour and this will undoubtedly make up your mind for you, for you will soon know whether or not you can put up with it.

One of the most important things to watch out for when you go to see a ferret, or a litter of ferrets, after the general health of the animals of course, is the environment in which they have been kept. The cage and the exercise area should be clean at all times. If the floor of the cage is yellowed with urine and there are piles of excrement built up, the animals stinking and, especially with albinos, their white fur stained yellow too, then my advice is to walk away. The hob in particular can have yellowed fur in places with sexual maturity, I am referring to staining by urine in a dirty cage. Diseases can soon rear their ugly heads where animals are kept in filthy conditions. In such a situation, it is easy to feel sorry for the animals but you must harden your heart and look elsewhere, even if the animals themselves, for the moment anyway, look healthy enough. The person with a clean and tidy home and of neat, clean appearance, will usually look after their animals well and will keep them in clean, well maintained cages, but do not be put off if you encounter a scruffy person whose house isn't the cleanest you have ever seen, for I have known people over the years who have thought much more of their animals than themselves and could not be faulted for their excellent animal husbandry, even though their personal hygiene leaves a lot to be desired.

Now that you are well equipped to go out and choose a suitable pet for yourself, or, if you are a parent, a suitable pet for your children, we must discuss the important issue of suitable housing for your ferret. It is important to get this right, but at least cages can easily be altered, or even changed completely, so this aspect of ferret care is a close second to the actual choosing of your pet. A ferret will be yours for the next few years so it is essential that you get this right the first time. It is unfair to the animal to be passing it on to someone else, or returning it to the breeder in exchange for another, so be patient and take your time choosing, for no one will mind you spending time playing with their ferrets before choosing one, if you choose one at all, as none may take your fancy and you may then need to look elsewhere. Rushing will only increase your chances of making a mistake.

Chapter Four

Housing

As soon as I acquired my first ferret I began to prepare a home for her. I built a cage from any scraps of wood I could find and some wire netting which I already had in the shed. The finished article was a real mismatch of around four feet in length and two feet in width, with no separate sleeping compartment, but at least it was secure enough. I settled it in a suitable spot in the back garden and scattered sawdust all over the floor of the cage and filled a corner with plenty of straw.

Satisfied that my ferret cage was good enough as a temporary residence, for I intended to save up and purchase a brand new one from our local pet shop – a place which fascinated me and where I spent much of my time as a youngster – I now went back to my friend's house and made my purchase.

I installed her in her new home and found a large piece of polythene sheeting which I placed over the cage to keep the wind and the rain off my new treasured possession. This was during the nineteen seventies and there were very few books about them on the market and, in fact, though I had tried, I could not get hold of one anywhere, so I was very new to ferret keeping and didn't really have a clue as to how to look after them properly. I had gained just a little knowledge from Sid's brief stay at our house and my experiences with him, but that knowledge was still very scanty and it is not at all surprising that my first encounter with ferret keeping for myself should end in disaster.

I grew up in a small two bedroom house which was part of quite a large estate, though where our house was situated was right on the edge of endless miles of beautiful countryside. Though beautiful in the height of summer with the hordes of house-martins flying to and fro from their nests which hung under the eaves of the house roofs, looking as though they would fall and spill their contents at any time, the place, being elevated and on the edge of the open countryside, was utterly exposed to the worst of weather conditions during midwinter and I can remember my brother and I scraping the pretty patterns made by the frost off the inside of our bedroom windows.

It was during the winter that my first ferret was purchased and housed in the back garden. I am sad to say, when I got up on that first morning and rushed outside with typical schoolboy excitement, impatient to see my pet again, expecting her to be at the mesh waiting to greet me, I found her dead, cold and as stiff as a board. Even though I had given her plenty of straw, the cold must have penetrated its protection and reached out with its cruel fingers and killed her. Of course, I do not know for sure that it was the cold that killed her, there may have been something else wrong with her, but I cannot help feeling that it was the inadequate shelter I had constructed for her.

Life is a learning curve and I never made the same mistake again. Having made better arrangements until I could purchase a suitable cage for my ferrets, I then set out on another mission of nagging and pleading and in the end was able to buy the second ferret from my friend's brother, who, by this time, was probably utterly sick of me and sold her to me just to get rid of the annoyance (whenever I wanted something, I used the same tried and tested methods on my mother, though, very often, she was a tough nut to crack and so I sometimes failed to get what I wanted).

This sad episode taught me that a ferret not only needs a cage to live in, but it also needs a nesting box to sleep in, a separate compartment where it can snuggle up and keep warm in even the coldest of conditions. Of course, in countries such as Canada where the temperature can drop rapidly, they may need even more protection from the elements. Caging a ferret in a Canadian winter outside may not be such a wise move. In the wild, animals which live in burrows will simply move deeper and deeper underground where the icy cold cannot penetrate and thus they are protected from even the worst of conditions (it is usually lack of food, rather than cold, which kills off many animals during a long, hard winter).

Even if a section of the cage is partitioned off in order to provide adequate sleeping quarters, I would still advise that a sleeping, or nesting, box be put into the compartment for extra protection. Filling this box with hay until you cannot get anymore into it, will give maximum shelter from the cold and ferrets will burrow into the hay and will make a little hollow deep inside where they will snuggle up and stay cosy and warm. If two ferrets share such a sleeping box, then the warmth is doubled from their body heat and they will be even safer during the winter of milder countries such as Britain and France. In cold, bitter conditions in countries such as Canada, and if you have decided that you do not want your ferrets in the home with you, then putting a cage in a suitable outhouse will help shelter them from cold, rain and snow, especially if you put a little heater inside the outhouse just to take the chill out of the air.

Whether you will be keeping your pet ferret in the home, or outside, either in the back garden or inside an outhouse such as a garden shed, you are going to need a cage. True, some prefer to have their animals living as cats do, free around the home and unconfined to a cage, but, until your pet is housetrained, or housebroken if you prefer, you will need to confine it and there are many excellent cages on the market that will serve this purpose. However, cages, good ones anyway, are not cheap and, if it is only going to be a temporary residence until your pet has learned to use a litter tray, rather than your floor coverings as a toilet, then you may not wish to spend your hard-earned cash on one, so, we must look for alternatives.

Some have chosen the bath as a temporary residence for their ferrets and this is something I would definitely not recommend. I, for one, wouldn't be too keen on having a bath in what, until a few minutes ago, was a ferret cage, so I do not believe this to be an option, though, if you do decide to use your bath, then keep the door shut as a ferret may soon catch on to any way of getting out of the tub. Baths aren't what they used to be. In the house I was born and and grew up in, our bath was huge and very deep and you could almost swim in it. A ferret would never have escaped from such a place, even a huge hob ferret, but the baths they make nowadays are much smaller and shallower too, so some ferrets may escape in time, after they have tried all different methods and made several attempts at escaping. Ferrets are great opportunists and will soon find a way out if there is one.

Large cardboard boxes may be suitable as temporary homes, but you must make sure that the box in question is made of thick card-board and is very sturdy. Also, you must be able to have a steady supply of such boxes for they are bound to get soiled at some time or other and will, as a result, soon become very smelly indeed. The odour of the ferret itself will also easily cling to this kind of material and this will not help to keep your home as ferret-odour-free as possible. Cardboard boxes are not at all ideal and their uses are very limited to say the least. Another problem with this kind of temporary home is that, if anything, or, indeed, anybody, were to fall on the box, then there would be very little, if any, protection for the inhabitant, and the result is both predictable and unthinkable – a crushed ferret.

If you have your mind set upon a cheap, even a free, temporary home for your pet, then I would suggest something a little more sturdy than a cardboard box. Instead, I would go for a wooden box made out of plywood which is very light and so is easily moved around when cleaning the home, and this type of box is also washable. These wooden boxes are generally used as packing cases and may be obtained from stores, or food suppliers, and they have the added attraction of being strong enough to give necessary protection to your ferret should an accident occur, for it is true that most accidents occur within the home and the safety of our animals, as it is with ourselves, is of the utmost importance.

One thing you must be careful of when using these wooden packing cases is protruding nails which could easily cause an injury to an inquisitive, highly active ferret, for these cases are made with little attention to detail and sometimes nails, or maybe large staples which can often be more dangerous than nails as they are much sharper, can be sticking out of the wood at dangerous angles. So check them over thoroughly before use and make sure that your ferret's home is a safe one. These cases are usually quite deep, so you need not fear your ferret jumping out and escaping, though, if the inside walls of the packing case are rough, there may be a chance of your ferret climbing out, so it may be advisable to put a little square of mesh over the top of it in order to prevent it from getting out. I think it unlikely that your ferret could escape, but it is always best to be safe, rather than sorry. The fact that a ferret will sometimes manage to climb up a rough piece of wood should alert us to the necessity of removing any sharp objects, even large splinters of wood, from the ferret's environment, for, if it slipped whilst climbing, it could very easily pierce itself and the consequences may be very serious indeed.

Even if you wish to allow your ferret the freedom of your home, I think it is still well worthwhile investing in a cage rather than messing about with temporary dwellings such as the ones mentioned, for there may be times when you want them out of the way, maybe when the mother-in-law, who is definitely rather anti-social when it comes to ferrets whom she considers smelly beasts, comes to pay you a visit! Also, in the future you may add more ferrets to your little family and they will need to be housetrained, so a cage will still continue to be useful long after your pet has caught on to the idea of using a litter tray.

I personally would not keep my ferrets permanently around the home, running loose. Ferrets are small animals and, despite their hardiness and downright grit, can easily be injured or even killed, if you were to accidentally stand on your pet. When cleaning or cooking, it would be so easy to become distracted and forget about your ferret as it wanders around, and one wrong step could cause a tragedy. Also, ferrets are incredibly active, quick animals, and could easily slip out of the front or back door without being seen and would undoubtedly become lost, possibly for good. If you are adamant about keeping your fitch indoors with you, then it is much simpler to keep it in a cage and then you can allow it some freedom around the home when it suits you. Even then, you must always be aware of the presence of the little fitch, lest the worst does happen. I say this from experience.

It wasn't a ferret that was involved in this incident, but it could easily have been. When we were considered old enough to carry the responsibility, my brother and I were allowed to keep our own pets, alongside the family dog and cats which were looked after mainly by our parents, and I chose mice while my brother went for gerbils. One of his favourites, Bunty, was out in the bedroom having a good run round with her mate, Benji. The bedroom door, for some reason, was slightly open and Bunty made a run for it. Mike, seeing what she was up to, tried to get in front of her to stop her getting out onto the landing where one of our cats may have been lurking, and, sadly, just caught her nose with his foot, killing her. I am certain she didn't suffer, but, nevertheless, it was still tragic, and a similar sort of thing could easily happen with your fitch, especially at a time when you are busy, so having them in a cage at such times when it is inconvenient for them to be out running free, is, I believe, the best method of keeping them, although, of course, you may disagree. In the end, it is entirely up to you whether you cage in the home, or have them living around the house.

I much prefer to keep my ferrets outside, not in a cage out in the open, for reasons we will discuss shortly, but in a cage inside an outhouse, or a garden shed. This is an excellent way of keeping ferrets. For one thing, it is then unnecessary to have your ferrets descented as they spend much of the time outdoors, only coming indoors when it is convenient. Also, if you make sure that the outhouse is both well maintained and secure, with all potential escape routes carefully and diligently blocked off, then your cage can be left open and the fitch can enjoy the freedom of the place, an extension, if you like, of its living quarters. It is simple enough to place a ramp from the floor to the cage, so that your pet can easily travel in and out, for, though ferrets are very active animals, they will wish to return, maybe quite frequently, to their nest box where they will enjoy a snooze, before returning to their hyperactive games. I cannot stress enough the need to make absolutely certain that any holes or cracks in the wood or brickwork of the building where you keep your ferrets are filled in completely, for ferrets can scratch at an opening and, especially if the wood is rotting, can soon make progress until, at last, an escape route has been created and then you may never see your beloved pet again.

It is the same in the home, if you will be allowing your fitch the run of the place, then carefully examine every room, every nook and cranny for possible escape routes, and then diligently fill them in, or get someone in to do it for you. Also, when you enter and leave your home, or whenever you answer the door to visitors, be aware of where your fitch is and make sure that it doesn't dash outside while you are not looking. If a ferret is allowed to escape, then it is in danger from dogs (though many dogs will not even attempt to tackle a ferret, especially one which still has its scent glands) and, more especially, from cars and other forms of traffic.

One thing to be aware of if you choose to cage your pet outside in, say, a wooden garden shed, is to make sure it is well ventilated, especially during the summer months when outbuildings can become very warm and stuffy, so it may be well worth having a couple of vents fitted (these should already be fitted in a stone, or brick building) and windows that open are ideal, though you must make sure that your ferrets cannot climb up to the windows and escape that way, for, be absolutely certain about it, if there is a way of escaping, your ferret will find it, and usually very quickly indeed, surprisingly quickly in fact. Don't forget, if they are going to succeed at making a living out in the wild, polecats, from which ferrets are descended, will explore every hole, however small, in their search for

43

a meal, and this inquisitiveness, essential to survival out in the wilds of the countryside, has been inherited by the domestic strains, adding to their already irresistibly charming qualities, but making it essential to keep a close eye on them at all times.

Types of Cage

The best cage for domestic ferrets is the welded mesh variety. The walls are best made out of one inch by one inch welded mesh as this means both security and cleanliness. Though very agile and capable of squeezing through some surprisingly narrow openings, a fitch would find it impossible to escape through one inch mesh. This material is very strong so there is no danger of your pet being able to make a small opening large enough to get through, for welded mesh is unyielding and a ferret can scratch and dig and chew all it wants, with no success whatsoever. Because of the holes in the floor, the droppings will fall through and, though a few traces of excrement may cling to the mesh in places, will remain very clean and odour-free too. True, a ferret may find it a little awkward walking on a mesh floor at first, but it will soon get used to it in time. The mesh for the floor should be of a smaller size or, except for the latrine, a covering can be put down.

The floor of the nesting compartment must be solid however, in order to keep out the cold when outside or in an outhouse, and the

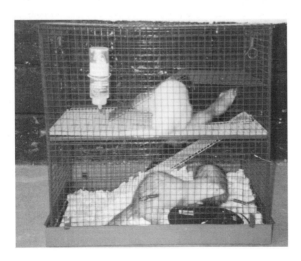

A two-tier ferret cage. Escape proof and roomy enough for a few ferrets, but this gives no protection from the cold, so it is only useful indoors or in a heated outhouse.

44

walls must also be solid, for the same reasons and also to keep draughts off the sleeping animal. Again, when a cage has a partitioned sleeping compartment, when outdoors especially, provide a snug nesting box too, for extra warmth. If it gets too hot, a ferret will simply lie outside its nesting box, but if it gets too cold, there is nothing it can then do to get warm and it may die, so make sure that it has both the protection of a sleeping compartment, and a nesting box placed inside the compartment.

This type of cage is mostly used by fur farmers as it allows the urine and excrement to fall through the mesh and so avoids soiling the fur of the mink. The cage is also very easily cleaned and does not harbour disease, or smells, so this type of arrangement may appeal to you. However, there are drawbacks to keeping ferrets in a cage of this sort and the worst drawback has got to be the cost. You may have to find an engineers or a metalwork shop and have one specially made, as I have never seen a cage of this type for sale in any pet shop,* and this will undoubtedly be very expensive. Even if you have a friend or a relative who is in the trade and they are willing to do you a favour, the materials themselves are rather expensive. Of course, the cost may not even merit consideration in your case and so you are determined to give your pet the best available accommodation. In that case, you must make sure that the cage is roomy enough for the number of ferrets you will be keeping in it.

A cage suitable for a maximum of two ferrets should be no less than fifty-one inches in length, sixteen inches in width and fifteen inches in height. Of course, if you increase the width, then the length can be shorter. I once saw a very good cage which was only thirty-seven inches in length, but was twenty-three inches wide, a very good alternative if you have a space for a cage which isn't very big in length, but is very roomy when it comes to width.

---

* There may be some pet shops which stock mesh cages and there might just be one in your area. If not, there are suppliers of similar type cages to be found on the internet, the websites being listed in the back of this book. Some of these are two or three level structures and this gives plenty of room for your ferrets. They are also very easy to clean as they have a removable bottom tray, or pan. These, however, do not come cheap. The main advantage of these is that they give much room for your pets, but do not take up too much space, making them ideal for use in the home. I would not recommend them for outdoor use however, as they provide no protection from the elements. There is a shipping service provided for overseas orders.

Though this cage is not very long, it is wide enough to give plenty of room to two ferrets.

You will have to come up with your own measurements according to the amount of space available, but the above measurements are probably the minimum for the housing requirements of ferrets which are, don't forget, very active and energetic little creatures which need space if they are to be happy, even if they will not be spending much time in it.

A welded mesh cage of these proportions will be well worth the expense, for the security and cleanliness of such a cage is of the very best available to you. The cage should be elevated on legs, with a sliding tray fitted underneath and the cleaning out of such a cage is made very simple indeed and it will take very little effort a couple of times a day, or maybe just once a day if you cannot spare the time, to keep it clean and free from smelly odours.

If you do not have plenty of finances available to spend on housing your pet, then purchasing a wooden cage is a much better alternative, and these have the added attraction of being readily available as secondhand objects for sale, which can easily be located in any local newspaper or journal that deals with such ads, though you must take precautions when buying a secondhand cage. First of all, check that it isn't damaged in any way, maybe a piece of wood has been knocked off, or possibly chewed off by a previous occupant, leaving a small hole which could easily be made larger by a determined fitch full of mischief. Secondly, make absolutely certain that the wood is not rotten anywhere. You can easily check for rot by gently poking at the wood in various places with a pen or a similar

This type of cage is ideal for outside the home, or in a garden shed.

object and, if the wood is soft and spongy, then leave well alone for a ferret will undoubtedly find a weak spot and will work at it until it can get free.

It is far better to buy a new wooden cage from a pet shop or some other outlet such as a garden centre or a hardware store, if you are in any doubt about the secondhand cages you have seen. When inspecting these new cages which are generally very well made, make sure that it is secure everywhere and that the doors are well fitted. If a door is a little out of line with the framework of the cage body, then a gap may be left which will attract the unwanted attention of any self-respecting ferret who wishes to uphold the family honour, so it is essential that doors fit properly and that the cage itself is well made and sturdy, with a good, strong mesh on the door, not a thin one which may rust or give when chewed constantly. A rusty mesh on a secondhand cage is also a potential escape route to a fitch, so avoid such, unless the cage is very cheap and you can replace the rusty mesh with new.

It is true that wood can hold odours for quite a while after a ferret has occupied it, but this isn't too much of a concern for the owner who is diligent about keeping their pet, and their pet's environment, clean, especially when the fitch has been descented.

If you live in a small apartment or flat and room is very limited, then a smaller cage, though not really recommended, may be an absolute necessity. A cage of forty inches in length, sixteen inches in width and fifteen inches in height would no doubt fit well into any small dwelling and, as long as the ferrets were given plenty of

Another cage well
suited to outdoor use.

exercise, would suit up to two of them, though you must be very diligent when it comes to keeping the said cage clean, for ferrets are
clean animals and it does them no good to be in dirty surroundings.
They will pick a corner as far away from their sleeping box as
possible, where they will deposit their droppings, so a small cage will
mean that they are much nearer to their toilet area than they would
choose to be under normal circumstances. It can only help in such a
situation to remove those droppings as soon as possible on a regular
basis, certainly no less than once a day, though it would be much
better if it was done morning and evening every day. This will not
only ensure that your pet is enjoying a healthier environment, but
will also help greatly to keep odours down to a minimum – something that is essential, especially in small spaces.

The ferret family is a clean family. True, polecats and some of the
other mustelids, will often take their prey into their bed chamber and
eat it there, afterwards sleeping on the bones and fur or feather that
are left over, but, unless the animal is sick, old, or on the brink of
imminent death, they will never soil their home and will usually
travel away from their chosen lair in order to do their business. This
is especially so with the European badger who is a very clean animal
indeed.

Badgers not only change their bedding on a regular basis, but,
very often, they will drag it outside of their setts where they will leave
it to air and freshen up, usually during a warm sunny day. This serves
two purposes. Firstly, this freshens up the bedding and gets rid of a
lot of the dust which can irritate a resting brock (an old English name
for a badger who is often referred to as Bill Brock) and disturb his
much needed rest. Sleeping is one of Bill Brock's favourite pastimes

and is something he takes part in with great eagerness, so he likes to freshen his bedding quite frequently. Secondly, leaving it out in the fresh air in this way will disperse parasites such as fleas and so, especially during a hot summer when fleas will breed and increase into plague proportions, he will do this on a more regular basis, for, again, nibbling parasites will disturb his slumber and so he is very keen to be rid of them.

So you can see what a clean family your ferret belongs to and I am sure you will understand the need to keep its cage and its nest box clean. I do not mean, of course, that you need to clean out the whole of the cage and nest box twice a day, not at all, for it is only necessary to remove the droppings frequently. Like badgers, ferrets prefer to have their toilet well away from home. If you ever get the chance to visit a well established badger sett, there you will find well worn paths leading to and from the sett, some to favourite feeding grounds, others to toilet areas known as latrines. True, some setts have been known to have rotting food, even badger droppings, inside them, but this is usually an old, or sick, or maybe an injured badger (many are hit by cars and are injured this way), and the rotting food may also be due to the presence of a fox somewhere in the sett.

We can learn much about our pets by studying the ways of wild creatures, thus helping us to give our animals, whatever they may

This ferret looks for a way out, but it will not find one in this well made cage.

49

be, the best environment and the best care possible. It is consideration itself on your part if you do your best to clean up droppings everyday.

At the time of writing, the cost of housing a couple of ferrets in a good sized cage is quite reasonable. I have just been looking at some very well-made cages at a pet shop and I thought their prices were very affordable. A cage fifty-one inches in length, sixteen inches in width and fifteen inches in height was on sale for £40.00 in the UK. The cage at thirty-seven inches in length, twenty-three inches in width and eighteen inches in height was on offer for £35.00, and the smaller cage which would suit small apartments, the one we have just discussed, was for sale at £39.00. Of course, prices will vary from supplier to supplier, from country to country. If you do not wish to be paying out these sort of prices, and if you do not wish to buy a secondhand cage which should cost you no more than half the price of a new one, then you could, if you have handyman skills, make your own, though this option may still be more expensive than buying secondhand.

The sleeping compartment to any cage should be no less than twelve inches in length (ideally it should be fourteen inches) with a width of around sixteen inches. Included in this compartment should be a nesting box of no less than eight inches by eight inches (ideally it should be ten by ten inches) and made of solid wood of around one inch thickness and with the joints very closely fitted so that there are no gaps to allow the cold in. An equally well fitted lid should lie on top to allow access for easy cleaning, and, of course, a

If you keep your fitches outside, or in an outhouse, make sure you provide a sleeping box placed inside the nesting area (the enclosed part) for extra protection from the cold.

small hole should be provided for access to the ferret. It need not be any deeper than ten inches. This will help hold the warmth, being a small space already inside a compartment, yet it will be big enough to hold plenty of bedding. Also, if you decide to breed at a later date, when you are more familiar with the ways of your pet, it will also be roomy enough for the kits.

Do not forget that ferrets are easily capable of squeezing through very small spaces and so it is essential that you check thoroughly that there are no gaps whatsoever in any area of the cage you are purchasing. It is also a good idea to check that any thin strips of wood are well secured as these may be pushed off by a ferret intent on exploring the outside world. I once converted a three tier bird cage into three separate ferret hutches, after I had failed miserably in my attempt to breed finches, and these had thin strips of wood fixed across a gap underneath the door, and one of these was obviously loose, for my ferret, a polecat type, or sable, named Jick, scratched and scratched at this loose piece until, finally, she managed to knock it off.

At that time my ferrets were kept outside in a shed and, during the summer when the shed could get a little on the warm side, I would leave the door wide open as well as the window, just to allow the cool breeze to keep the temperature down, and Jick, now free from the confines of her cage, simply walked out into the big wide world and disappeared into the darkening evening. A ferret on the loose is quite a worry. I feared both for her and any pet animals she came across, for she was only familiar with dogs (I had four at the time) and I was worried that she might attack a cat if she came across one, not to mention someone's pet rabbit feeding contentedly out on the grass (remember, ferrets are natural predators). Fortunately, after she had enjoyed quite an adventure wandering around all over the estate where I grew up, she was picked up by someone I knew and so she was safely returned, after my friend, Norman, had found her exploring his garden.

After that episode, an episode which could easily have seen my favourite fitch lost forever, I have alway checked, and then double checked, the sturdiness of my hutches.

Cages with doors which open from the top and come down from the front of the cage, or doors opening from the side are the best for housing ferrets. A hutch with the door opening from the bottom of the cage at the front is rather awkward, for, as soon as the door is open, your ferrets are out before you have finished opening it and they can then be all over the place before you know it. I would also advise that you never house a ferret in a cage with a door on the top of the cage for fairly obvious reasons. A lapse in concentration could end in tragedy. Ferrets are good climbers and will usually climb straight up the mesh as soon as they have realized you are opening the door, and will pop their heads out immediately they are able. If the door was to slip out of your hand as you were lifting it, then the result is easily predictable. A ferret wouldn't stand a chance, so the risk just isn't worth it, especially if a child will be doing most of the work in looking after it. Adults may take much more care with a cage of this kind, but still accidents do happen, so it is best to avoid a hutch with a top door. Admittedly, I have kept ferrets in this type of cage before and I did avoid any accidents, but I would not use this arrangement again, for it is not worth the worrying.

A door which drops from the top at the front of the cage allows you to keep your ferrets in control much better, as does the door opening from the side. The door which opens from the bottom is also a bit risky, as well as awkward, for the same reasons we have already discussed regarding the top door.

Having chosen the materials and the size of the hutch which best suits both your bank balance and the room available to you, and having settled upon a place for your ferrets, either in the home or outside, it is time to purchase your cage and then get everything set in readiness for the eagerly awaited arrival of your pet. If you are having your ferret caged outside, rather than in an outhouse such as a garden shed, then you must be aware of a few drawbacks to keeping small animals in this manner.

Firstly, you must make absolutely sure that the cage is covered with a waterproof top and, even then, you must also keep it covered with a good, strong waterproof sheet to help keep both the wind and the rain from getting to your ferret. Secondly, you must be aware of the need for a very sturdy cage in order to keep thieves out, for ferrets are very popular and anything with even a small measure of popularity soon becomes a target for thieves, so fit the cage with a good strong lock and make sure the mesh is thick and hard to break, for a thief will soon get through thin, flimsy wire netting. Of course, with pet rabbits kept out in the garden, in some countries such as Britain and Ireland, there is always a chance of foxes tearing the mesh, or breaking the door to get at them, but this

Ferrets love their comfort. A hammock is ideal if you keep your fitches indoors, but this type of bed will not give enough protection from the cold when they are kept outside the home.

is unlikely when it comes to ferrets, for foxes will not usually tackle them. Many towns and cities are now full of foxes, but you needn't worry too much about these. It is the stray dog which poses more of a threat to your ferret caged in this sort of situation, so, if this is the only option available to you, then please make sure that the cage is well built and is very sturdy, for stray dogs will soon make their way into a cage with any sort of weakness.

It is also important that a wooden cage be treated with wood preservative if it is going to be standing out in all weathers, but do not treat the inside of the cage at all, this should be left as bare wood. A ferret scratching and chewing at treated wood may harm itself.

Essential Equipment

You will need a water bottle for the cage and these are easily attached to the mesh front with a piece of thin wire which comes with the bottle. Water bottles are by far the best method of supplying fresh, clean water to your fitch, as dishes of water are liable to be knocked over, or the water is easily soiled by ferrets kicking debris into it as they pass.

A food dish is the next item you will need *before* you go out and purchase your new pet. A weighted dish is best as many ferrets will easily overturn a lightweight dish and will spill the contents all over the place, doing nothing for the cleanliness of the cage, for some kinds of food will begin to smell if they are left on the bottom of the hutch, undetected because of being mixed in with the sawdust on the floor of their home.

A litter tray, or litter pan as they are known in America, is essential if you wish to keep the hutch both clean and relatively odour free. This should have sides just a little taller than the ferret so that the droppings do not spill over onto the wood, and is best if it is made of plastic, for it will then be easily washable and, more importantly, will not absorb odours. I believe that litter trays are now available exclusively for ferret use and some of these are made to fit in corners, which is ideal as most ferrets will back into a corner when excreting. Also, they are designed so that your ferret can easily back into them. It may be well worth getting in touch with some of the suppliers listed in the back of this book, and ordering one of these specially designed litter pans for your cage, and, when housebreaking, for around the house.

For both the litter trays and the bottom of the cage, I find sawdust to be the best material to use. Pet shops supply treated sawdust which is extremely clean, but it is not cheap. A better alternative is to go to a sawmill or any place which supplies wood and ask if you can have the shavings and the wood dust from around the place. There may be a small charge for such materials, but, generally, you can get a sackful of sawdust and shavings for next to nothing, possibly free of charge as it saves them time cleaning up. I always went to our local sawmill when I was a lad for the shavings for my ferrets and usually got a huge sackful free of charge.

For materials for use in the nesting, or sleeping box, you could use old towels or sheets cut up into smaller pieces, but these will need changing and washing very frequently as odours will cling to them in no time at all. You may wish to try shredded paper which I am sure you could get hold of from local offices, though the print, especially on an albino, may come off on the fur and spoil its appearance, something which you would want to avoid, especially if you will be showing your ferrets. This material is one which I haven't tried, but it may well be worth experimenting with, though, of course, you can buy hamster bedding material which is often shredded paper, but is not a cheap option.

The very best method of keeping your costs down and providing your pet with a very warm and comfortable bedding material, is to use hay or straw purchased from a farm or a stables, usually very cheaply. A bale will last an age and is an extremely economical way of supplying necessary bedding material. True, this sort of bedding is readily available in pet shops, but you will have to pay quite a high price for it when compared to buying from a farm or a stables, though, of course, what you get from the pet shop is very clean and of the highest quality. It is possible that farm bought hay or straw may not be of the best quality, but I have never had any problems whatsoever using this method and, for ferrets kept out of doors in particular, it is extremely warm material. Ferrets will burrow into hay especially, my more favoured type of bedding, and will make a nest right in the centre of a thick tangle of it, which you have stuffed into the box, keeping your pet safe from the cold outside.

We now have our cage, maybe made of welded mesh, or a standard wooden one, we have situated it either indoors, out in the garden, inside an outhouse, a wooden hut or maybe a garden shed, and we have made sure that it is secure, that the ferret cannot escape, and, just as important for those kept out in the garden, that burglars or other animals cannot get into it. We have our litter pan or tray at the far side of the cage, well away from the sleeping area, sawdust is scattered on the cage floor, soaking up any moisture or dirt, and the sleeping box is full of suitable material. The bottle is full of fresh water and the food dish lies empty as yet, though that is about to change.

Having at last got set up, our ferret has been chosen from a reputable and reliable source and is now installed in its new home. A good feed will now help it to settle down a little in the unfamiliar surroundings, but, before we discuss the feeding of ferrets, we will consider their handling, for this is one of the most important aspects of ferret keeping and their husbandry.

Chapter Five

Handling

The internationally screened television show, *It'll Be Alright On The Night*, is well known for portraying often hilarious out-takes from film and television and one of the best known of these is an early clip of Richard Whiteley, better known for presenting *Countdown* on British television, being bitten by one of Brian's Plummer's ferrets. He was appearing on television to promote his book on ferrets and ferreting and, if I am not mistaken, he appeared alongside three of the subjects they were discussing that day, one of which promptly sunk its teeth into the presenter's finger. As if this wasn't entertaining enough, Brian Plummer then asked Richard Whiteley to hold the other two ferrets while he attempted to get the other one to loosen its grip!

This illustrates that ferrets can bite when in a situation which makes them feel threatened, for I am certain that the unfamiliar surroundings, the bright studio lights and the large number of people around them, made them feel this way and so one of them resorted to biting as a natural defence mechanism. Either that or Richard Whiteley smelt strongly of rabbits!

Knowing that ferrets can resort to biting will help the owner to show their pet the respect they deserve. To claim that ferrets are vicious however, is totally unjustified, but it remains an image that the public in general continues to have of these beautiful creatures. Everytime I have mentioned to people that I am writing a book about pet ferrets, I have had the same reaction. 'Oooh, they're vicious little things, aren't they?' is what they say and I always go on the defensive whenever I hear this. Most people have never had any contact with ferrets whatsoever, yet their first association with them whenever they are mentioned is of a smelly, vicious animal. I blame ill-informed television programme makers for this, as this is usually how they are portrayed.

The television series *All Creatures Great And Small* based on the

Charlie Hitchen, aged eleven, with Sox.

experiences of Yorkshire vet James Herriot once featured a ferret. The animal was brought to the surgery because it wasn't doing its job of hunting rabbits properly. The advice given was to stop the children from handling it and to starve it before taking it out hunting. This is the worst advice that could have been dished out. Firstly, a well handled ferret which is given plenty of attention, particularly by children, will be a delight to own, whether one works it or keeps it as a pet, and secondly, the truth of the matter is that a ferreting man will actually feed his animals before hunting them, in order to avoid the ferret which is sent below ground, from eating the rabbit which it may kill inside the warren, should it refuse, or be unable to bolt.

Young ferrets of six or eight weeks of age, the usual age when kits are sold, will have a tendency to nip or even bite. This is known as 'mouthing' and is the same kind of behaviour observed in puppies and kittens. This is very natural and you must not be put off from purchasing a kit which 'mouths' your fingers, for it is only doing what comes naturally. We have to take another look at the wild polecat in order to better understand this behaviour which can so easily be misunderstood and can often put people off keeping them.

Young polecats need to learn important lessons in life very early if they are to survive for very long once they have left the care and the protection of their mother who will be extremely diligent when raising her litter. The most important lesson polecat kits need to learn is what is, and what is not, acceptable as a food source, and that, basically, is why ferret kits are so prone to 'mouth' your fingers during the early stage of their lives. They will also nip you whenever you play with them and, again, this is only natural, for play teaches young predators two very important lessons. One, how to defend themselves against attack, and, two, how to kill their prey.

However, for the pet ferret, these lessons are not so vital, so you must quickly teach your young pet that, though gentle 'mouthing' is acceptable during play, applying pressure and even going mad to bite you is not acceptable at all and that this behaviour will not be tolerated. These are lessons which will soon be learned, provided you spend time with your new pet and handle it on a regular basis. This is the basic way of ensuring that your ferret will not grow up to be a biter. Plenty of contact with their human companions is absolutely essential.

Ferret kits, don't forget, are used to competing for food with their litter mates and will often grab at whatever they assume to be food. This may be your finger, or it may be that a kit just gets too rough while enjoying play, play which would normally be enjoyed with their siblings who will be more than happy to cope with a little rough treatment, for their fur will give much protection and will greatly soften a bite. Our skin is soft and sensitive and will feel it much more. So what can you do to stop a kit from 'mouthing' too hard?

As I have already said, the first thing to do is to handle your young pet on a regular basis and in a gentle manner, for rough handling will do nothing to help the situation and may actually make it worse. Having said that, ferrets are tough little creatures and should not be handled as though they are delicate china. The second thing to do is to make sure that you handle your pet in the correct manner. Your

Ferrets are easily handled, making great pets.

thumb and forefinger should be placed around the neck, while the other three fingers go just behind the front legs. Some advise that you should support the ferret by putting your other hand under its back legs, though this is unnecessary, but, if it makes you feel better, then there is no harm in doing this.

Never, ever pick a ferret up by the tail, as this could cause injury and would also do nothing for its temperament. If anything would induce a normally placid ferret to bite, it would be because of its owner picking it up in such a way. And do not use the scruff of the

59

neck either. This way of handling a youngster will not harm it at all, but it is not getting the proper amount of contact, for picking it up in this manner is as though you are keeping it at a distance. Close, personal handling is always the most effective method which will get results, leaving you with an easily handled family pet, one that can be trusted with children and adults alike.

The third thing to do in order to prevent your kit from growing up a biter, is to administer discipline whenever it 'mouths' you too hard. Giving your ferret a gentle spank on its rear flank, while firmly saying 'no', will help your fitch to decipher just where it is going wrong, but this must be done while it is biting too hard, so that it learns to associate the punishment with the misdemeanour. A gentle tap on the nose will also work as effectively and this is the method I generally use, but I must emphasize the word *gentle* when it comes to spanking. Remember, that although ferrets are hardy creatures, they are only small compared even to a child, so you must be very careful not to injure your fitch. If a child will be doing most of the work and care of the animal, then you must teach that child both how to handle and how to administer discipline in the correct way.

Ferret kits bought from a supplier who breeds ferrets in quite large numbers, from a pet shop for example, will probably have had very little handling during the period between weaning and the time for them to be sold, so these will be more prone to 'mouthing' than kits bought from a private breeder who will usually breed in small numbers and so will be able to spend much more time with their youngsters, thus having them well used to being handled by the time they are sold. As youngsters though, even these will be 'mouthing' the fingers of their new owner and will need to be handled just as regularly as those with little experience of human contact. I always handle and play with my young ferrets, long before selling them on.

Whenever you pick up your new pet, it may wriggle and squirm and get up to all manner of tricks in order to free itself. This is natural and you must not respond by squeezing hard in order to hang onto it. Simply swap hands and keep doing this until it gets fed up and stops struggling. It will soon get acquainted with all of these new experiences and will learn to take them in its stride in time.

As you handle your young fitch, you may wish to reward it with small treats which can be purchased at most pet shops, or from any of the internet suppliers, but I simply reward a ferret with plenty of stroking and especially by tickling just behind the ears which they enjoy immensely. Most ferrets' hard exterior will simply melt away when you gently rub them behind their ears and many will curl up and go to sleep in your lap. This process not only teaches them that being handled is a very pleasant experience, but it is also doing much to socialize them and get them familiar with human company. This in turn will produce a non-biting ferret which will be pleasant company for many years to come. And, even if you do not have children yourself, get a niece or a nephew round to play with your ferret and so get it used to being handled by other people as well as yourself. If you are a youngster, then why not invite your school friends round and have them handle your fitch, though you must make sure that they are not too rough with it as this may upset it and spoil the handling process which you have put so much effort into accomplishing.

Though I do not advise that you handle a young ferret by its scruff, the loose skin at the back of the neck which the mother will use for picking up her babies, something that is totally unnecessary with a kit, it may be that you will have to use the scruff of the neck if you have chosen to keep an adult that is now showing a tendency to bite. Holding a ferret in this way will prevent it from biting you and you can hold it in this manner while you firmly tell it off for biting or even attempting to bite. With an adult biter, as opposed to the youngster, you could receive a nasty wound and, believe me, the bite will hurt, so wear a protective glove on the hand you normally use for holding your pet, making sure first of all that you rub your scent all over the glove, and use your free hand for grabbing it by the scruff of the neck everytime it bites, or attempts to bite, telling it off in a firm tone of voice. Repeat this procedure a few times, but not too many times, and either put it back in its cage or allow it to have a run round for a while, long before it becomes fed up and so will grow to hate these essential handling sessions. Believe me, in time you will not need that glove anymore and your efforts will have been well worth your while, though you must be patient and you must persevere.

If, however, after quite a long time of these gentle handling sessions, the 'scruffing' and the telling off, along with rewards for good behaviour, your fitch continues to bite, then you may wish to pass it on to a more experienced handler who might just be able to cure it of this bad habit. If not, then do not be tempted to release it out into the wild as it will undoubtedly starve to death. True, a

Stephanie Burke with Pippin. A ten-year-old girl can safely handle a well socialized fitch.

ferret which is used for the hunting of rats and/or rabbits, will be able to survive in areas where their prey is found in suitable numbers, but a pet ferret would have no idea of how to fend for itself. Even a working ferret would have the odds stacked against it, so the non-worker would have no chance of survival. I would much rather have a persistent biter put down by a vet than release it to fend for itself, but only after all other avenues had been exhausted.

Most ferrets that have a tendency to bite can, with time and patience, be cured of this bad trait and release, or even putting them down, is totally unnecessary. Even Sid Vicious, if I had him now, would have been cured of his biting habit with much hard work and persistence, but this situation can easily be avoided by purchasing a kit, rather than an adult, until you become adept at handling. Or, if you are determined that it is an adult you most favour, by going to a reputable breeder, or to a ferret welfare, or rescue centre where you will be provided with ferrets that are friendly and of good disposition.

Just a brief word about where not to put your fingers when you are telling your ferret off for 'mouthing', or even biting. Do not wag your fingers in front of his face and do not put your face up to his, or hers, in a threatening manner, or you could easily find yourself bitten again. Be firm and consistent with discipline whenever your ferret nips or bites and it will soon learn to be more sociable, allowing you to build a much friendlier and more trusting relationship with your new pet.

I cannot stress enough the need for both patience and perseverance when it comes to getting animals well acquainted with human contact. It was a young fox which taught me this very valuable lesson many years ago. A litter of fox cubs had been captured from a local dump which stank and was rat infested, a place where many foxes will die slow, agonizing deaths from that terrible skin disease, mange. I obtained one of these cubs, a six week old vixen who, understandably, was incredibly shy of humans and avoided any contact with them, even eye contact. I named her Sly and set about fixing up a little home for her inside my garden shed where the ferrets didn't seem to mind the new occupant at all, though I made absolutely sure their cages were in good order lest the fox try to make a meal out of them. My cages were good and strong with welded mesh fronts, so I was content to have Sly in amongst them without any worries.

Whenever I entered the shed to fetch my ferrets in order to exercise them, clean out their cages, or feed them, Sly would head off into her shelter well out of the way and would not emerge again until I was out of her space. I would sneak around to the small window and take a peep inside, and from there I would watch her coming out of her cage and either taking a drink, or feeding, or just looking around the place with typical foxy curiosity, always alert, always on the lookout for danger, or possibly for food, and always shying away from me whenever she discovered my presence.

However, with patience and perseverance I kept at it until at long last she would emerge from her little hiding place and actually take food while I remained in her presence, though I always made sure that I stayed very still and quiet in order to avoid spooking her and once again sending her scurrying for cover. Sly soon got used to my presence at mealtimes and I then began offering her food from my hand. She was very reluctant at first, but I persisted and eventually she very slowly and carefully approached and cautiously accepted my offering, though she would take her food well away from me to eat it.

I then was able to progress to actually stroking her while she ate and then I moved on to spending time with her between mealtimes and she would allow me to stroke her without any hesitation after a while, until, after quite a long time and a lot of hard work, Sly accepted me fully and would even play games with me, hiding away and then running out and pretending to bite me, gently 'mouthing' my hand in her powerful jaws. She was a beautiful animal and I enjoyed keeping her, but there were drawbacks. Sometimes she would bark at night and she was very smelly, far worse than my ferrets which were not descented, for she was dirty of habits and I soon began to sympathize with badgers who will sometimes have to suffer a fox taking up residence inside their usually clean setts.

Sly became friendly and accepted my company very naturally after a lot of patient and persistent handling. If results of this kind can be obtained with a wild animal, how much more so can we succeed with our ferrets which are, after all, domesticated animals. So be patient, be persistent, be gentle and, above all, be regular in your handling of your new pet ferret.

Chapter Six

Feeding

When I first began keeping ferrets the food available was very limited indeed and their tastes were not catered for at all in the pet food industry. Ferret feeding back then was more of a lucky guess, based mainly upon the diet of the wild polecat, than guided animal husbandry.

I grew up in an area bordering endless miles of countryside and the fields and woodlands were literally full of rabbits, so quite a few of my friends caught wild rabbits which were a part of their diet, so I would be given the inedible parts which made ideal ferret food. Ferrets love rabbit heads in particular and this kind of diet not only produced the protein they need, but also, because of the bone and fur, produced plenty of roughage.

I worked on a chicken farm for a while during the early days of my time spent with ferrets, and I was allowed to take any dead chickens home and these made excellent food for them, though, of course, this type of food is only useful for ferrets kept outside as bits of meat, feather, bone, or fur, are not very pleasant to have around the home so unless you house your pet outside or in a shed of some kind, I would not recommend this type of feed. Also, if your ferret is not descented, for some extraordinary reason raw meat will increase the smelliness of your pet and this, of course, is not desirable when you have them either living indoors and sharing your home, or enjoying exercise around the home. Their droppings too, will be a little smellier when they are fed on a diet of raw meat, though, obviously, it is the most natural way of feeding them. Working ferrets are often fed on those parts of the rabbit which are not suitable for human consumption and this is an extremely cheap way of feeding your animals, for it will cost you nothing, though, even if you have a plentiful supply of rabbit meat, varying the diet a little will be of benefit to your fitches, for, as the saying goes, 'variety is the spice of life'.

Nowadays, of course, things have moved on a great deal and there are plenty of choices for the ferret owner who is diligent about providing his/her pet with a healthy, balanced diet, one which will provide all of the essential ingredients in order to keep their animals in peak condition. Remember, your ferret is a predator, a natural meat eater and so a high protein diet is essential if it is to remain in good health. Also, ferrets are highly active animals and will burn energy at a rapid rate, so they will usually eat quite a lot of food for their small size during an average day. Make sure that you provide a diet with plenty of protein included, that means a diet which contains meat, and make sure also that there is plenty of it. Ferrets will very often eat a portion of their food and then return a little later for more, usually at regular intervals throughout the day, so make sure there is food in the dish for most of the time, along with a plentiful supply of clean, fresh water. A large water bottle designed primarily for rabbits and guinea pigs is ideal for this.

The feeding of live prey to ferrets is totally unnecessary, for ferrets do not eat only after killing their prey for themselves. They are carrion eaters, that is, they will eat dead meat that they come across, not live prey only, so offering live mice to your ferret is not at all the way to go and this serves no useful purpose whatsoever. It may even bring a prosecution for cruelty from an animal welfare organization, as ferrets, unlike snakes and spiders, do not need to be presented with live prey in order to survive.

Carrion eaters will often 'cache', or store their food and return to it later. Do not try to stop your fitch from doing this as it is only obeying its natural instincts. In the wild, polecats will store many pieces of leftover food and this serves a very useful purpose. If a polecat kills a fully grown rabbit, or a large bird such as a chicken from a farmyard, it will be unable to eat all of its prey at one sitting, so it will seek out suitable hiding places, such as deep inside a rabbit hole, under large slabs of rocks, or within a twisted mass of tree roots, and will leave it there. In some cases it may even dig a bit of a hole and bury its food there, hoping that it will not be discovered. The polecat's instinct to 'cache' is an act of survival as, particularly during periods when food is a little scarce, it will return to these little 'storehouses' and will eat the remains of a kill it may have made days before. Very often though, especially where food is abundant, particularly during summer and autumn when prey species are at their highest numbers, they will forget about the food they have put aside for leaner times and will leave it there to rot. Foxes also 'cache' their prey and, like polecats, will forget about it.

This natural survival instinct can cause the ferret owner a few problems. If you keep your ferrets indoors, either caged or running loose as some do, then you do not want to be feeding raw meat in particular as you may begin to discern unpleasant smells as the meat begins to rot in places you probably didn't even know existed, so, in this situation you need to be careful about what you feed your fitches. If you find that your pet does indeed 'cache' its food, do not try to prevent it for you will only wear yourself out chasing it around the place, and, anyway, in time, once your ferret has realized that there is a plentiful supply of food, it may lose interest in storing bits of food and this behaviour may subside altogether, though this is doubtful I have to say. Even though they may stop storing food all over the place, you will find that they continue to hide it in their bedding box, hence the need to clean both the cage and the nesting box on a regular basis, both for the sake of the health of your fitch, and in order to keep odours to a minimum. This is especially important if you use sawdust for the bottom of the cage as your fitch may bury food under it, usually in a corner somewhere.

Some choose to feed their fitches on a good quality dry cat food and this, if you desire to feed this sort of diet, is suitable as it will be a good, balanced and nutritional feed, though every few days it is best to feed them something different for variety so either canned cat, or dog food will do. Of course, you could feed your fitch canned food most of the time, but you will need to supplement this with a higher protein feed. I very often feed my ferrets on either canned cat or dog food, supplemented by rabbit heads. The only problem with feeding canned feed is that your ferret's droppings will be quite smelly, though, if you are diligent about cleaning, this will not be too much of a problem.

Dry foods made exclusively for fitch food are widely available and, thankfully, are not too expensive to make them an option for those of modest means, though some brands are more expensive than others. These feeds are excellent and come highly recommended, for they are well balanced and have all of the vital ingredients needed to keep your pet in good health, for the vital ingredients are supplied in the correct quantities.

Essential Requirements for Feeding

A well balanced feed should provide your fitch with the following vital ingredients which will help it to live a long, active and healthy life. *Water*, which is, of course, absolutely essential, though you must always make sure that there is also plenty of fresh, clean water available at all times. This can be supplied in a water bottle attached to the mesh of the cage, and in a heavy pot dish when your ferrets are loose around the home, making sure that the dish is heavy enough to prevent it from being knocked over by a mischievous fitch. The dish should also be shallow so that your pet can get to the water much easier, and to avoid a drowning, especially when young kits are running loose, for these, like young puppies or kittens, can easily succumb to fatal accidents if you are not careful, so avoid large, deep water dishes.

Protein, for building muscle and also for repairing muscle and tissue. Ferrets are highly active creatures and they require quite large amounts of this ingredient. Though canned meat is a little low in protein, dry ferret foods have a high protein content that is well suited for your pet's needs. This will help the growth of muscle and will aid repair after strenuous use, something that is essential for highly active animals.

Fat content in ferret food will help them to keep warm and, of course, fat is turned into energy, so quite a high fat content in the diet is absolutely essential, otherwise your ferrets would be lethargic. Too much fat, however, and your pet will become obese and will suffer health problems and undoubtedly an early death. The specialized ferret foods have the correct amount of fat included in the ingredients, so all of the hard work is done for us.

Phosphorous and calcium aid the growth and maintenance of strong and healthy teeth and bones.

Fibre, or roughage, which will help keep the digestive system working properly, keeping your ferret regular.

Salt will also be included in a complete ferret food of quality, but only in very small doses, for this mineral is essential for regulating the proper balance of water inside the body, though this would be harmful in large doses.

Essential vitamins such as vitamins A, E and D should also be included in a quality complete food, but, again, in correct doses as an overload of vitamins, or, indeed, a lack of them, can be very harmful.

I strongly advise that you feed your ferret on one of these brands of complete ferret food and one of the brands which has earned a very good reputation in recent years is James Wellbeloved Ferret Complete which can be bought in both two kilogram bags and twenty kilogram sacks, making it ideal for those who keep only one or two ferrets, or a whole host of them.

This brand contains meat, poultry, fish and chicken gravy which gives it a protein content of thirty-six percent, a very good amount indeed, far more than dog and cat food will supply, thus making it much more suited as food for your fitch. Unlike raw meat or even canned meat, this product will remain fresh and will not smell and thus attract flies, or negative comments from visitors.

Because ferrets will not eat to satisfaction at one sitting, returning time and time again to their food bowls throughout the day, this quality of remaining fresh for upwards of a couple of days in the bowl, makes it ideal. This will also prevent stomach upsets which can easily result from raw meat being left in the bowl for too long. Another benefit is that a bowl containing food throughout the day may discourage your fitch from storing some of its food in little hiding places, thus the cage and your home will remain much cleaner and, just as important, odours will be kept to a minimum. Though, even if your fitch does continue to store some of its food, because it is a dry food which will remain fresh for much longer, you do not have to worry about rotting meat, maggots, or bad smells.

Ferret Complete is rich in full fat linseed oil which will help to keep the skin and fur healthy. The coat should be glossy and these essential oils will contribute greatly to a healthy, shiny coat, something that is important at all times, but especially for the ferret which is exhibited at shows. A poor, lifeless, dull coat is both unsightly and shows that there is something wrong with the ferret. These oils are absorbed by the coat and this will really aid the natural nourishment of the fur.

Because this product contains poultry, fish and plenty of rich chicken gravy, it is very tasty for ferrets and this, of course, makes it an ideal feed for your pet, for dry foods which are like rabbit foods, will not go down well with the average self respecting fitch who will want a good meaty flavour every time.

With the average daily cost of feeding one ferret very low, this kind of feed is both good for your pet, and good for your purse. Also, this dry feed, because it is well balanced, will be thoroughly absorbed and so there will be far less waste, thus less smells from the droppings, something which cannot fail to appeal, especially to those who keep their ferrets indoors.

There are other brands of dry complete ferret foods on the market today, now that ferrets are getting the recognition and the attention they deserve. Frankie Ferret Complete food is another brand which contains chicken and turkey and essential proteins, vitamins and minerals, with linseed oil for the goodness of the coat. Again, because they contain real meat, these dry foods are tasty and Frankie Ferret boast that their feed is easily digestible as it contains plenty of fibre. The recommended dose is between fifty to sixty grams per day, per ferret, depending, of course, on the size of the animal. Both of these brands are readily available, though, if you have any trouble obtaining a suitable complete feed from your local pet store, then try some of the suppliers listed at the back of this book.

For those who keep their fitch indoors, allowing them great freedom around the home, whether they are caged or not, then, as I have said, the feeding of dry complete feeds is by far the best option available, but that is not to say that you cannot, on occasion, feed your ferrets on other suitable foods. Eggs, for instance, are very good for them and a fitch will thoroughly enjoy feasting on an egg,

Complete ferret food comes highly recommended – James Wellbeloved also supply a high quality complete meal.

though I would not recommend feeding them eggs too often. In the wild, polecats will raid the nests of wild birds, or even nests of chickens inside a coop, and will eat the eggs with much enthusiasm to say the least, for they are no doubt a welcome change after an endless diet of meat, blood and guts. They will enjoy both the tastiness and the goodness of eggs, as, indeed, we humans do, but, you must remember that eggs, to a wild animal, are a seasonal feed and are only available at certain times of the year, unless, of course, a farmer is negligent enough to allow a roaming polecat into his hen coop whenever it felt like it, which is very doubtful indeed. This rich food source is there for the taking during the spring and early summer months (some species will lay two clutches during the nesting season), so, at all other times of year, except, as stated, when one raids a hen house, polecats will not have eggs included on their menu. The conclusion is that you must not feed eggs to your ferrets on too regular a basis, as you may do far more harm than good. You could either mimic the seasonal feed by only providing eggs say once or twice a week during the natural nesting season (wild polecats will not find nests to raid every single day of the nesting period), or you could feed eggs as a treat maybe once a fortnight throughout the year. Of course, you may wish to give them eggs on a more regular basis. If you do, then be careful and take note of their droppings. If they become too loose and very smelly, then it is probably because you are feeding them eggs, or possibly some other treat, far too often, so cut down on these treats, at least by half, though you would be better cutting them out altogether for a few days.

Some, especially those who keep their ferrets outside, either in the garden in a good sturdy cage which will keep both predators and thieves from getting to your fitches, or inside an outhouse, possibly having the run of the place, choose to feed day old chicks. These are quite widely available and are ordered in large numbers which can be kept fresh by freezing them. At one time these day old chicks were the main part of a ferret's diet and, indeed, many were fed exclusively on them. Because research indicates that there isn't enough of the essential ingredients found in this food source, most ferret owners are supplementing this diet with other things, to make it more balanced and more beneficial. Day old chicks are, of course, a good source of fibre, but you must include other things too, such as canned dog or cat food, raw meat such as liver, and eggs (they love hard-boiled eggs).

Ferrets will often eat scraps from the table too, such as left-over chicken or lamb, or maybe the innards which often come with a chicken, but be careful not to feed your ferret anything of this kind along with the bones.

If you work your ferrets, or you know of someone who does, then you will be fortunate enough to be able to feed rabbits to your fitches, giving them everything to feast upon (you may be best removing the intestines which could contain parasites, and the liver if it has white dots on it, indicating that it is diseased), the bones included, for ferrets will eat raw bone which will do them no harm, but cooked bone changes in structure and splinters easily, splinters which can do great damage and even kill your fitch, so *never* feed cooked bone at any time. This may seem a waste to you, all of that chicken carcass going to waste, but you can use it. Boil the carcass in a pan and make, with a chicken gravy cube added, a rich gravy which you can pour (after all bones have been sifted out of course) onto your ferret's food, adding greatly to the flavour and enriching its feed with more goodness.

Unbelieveably, some ferret owners have striven to feed bread and milk to their pets as their staple diet and this, as you can imagine, will lead to a very unhealthy and unhappy fitch. Ferrets need meat as the staple part of their diet and ferrets must have meat if they are to thrive. Those who choose to keep their fitch on a diet of this kind should be prosecuted for cruelty, as they are harming their pet, rather than looking after it properly. Having said this, however, I believe there is nothing wrong with feeding bread and milk to a fitch, as long as it is only on occasion, maybe when you have been too busy to notice that you have run out of your normal feed and cannot restock until the next day. If you do choose to give bread and milk occasionally, as I have done many times, then make sure that you water down the milk with at least one third of water to two thirds of milk, for milk is a little rich and may give your fitch diarrhoea. It is also best to use wholemeal bread, for there is precious little good-ness in white bread. I must stress, however, that you only give bread and milk *occasionally*, as a treat for your fitch who will relish this tasty dish.

Fish, too, can be given as a treat on occasion, but I would not feed this too often, mainly because of the smell of fish, but also because fish will make up only a very small part of a wild polecat's diet which will live mainly on rabbits, rodents and birds. If you do feed fish, then remove the bones first, as there is always the danger of a bone becoming lodged in the throat. This is most unlikely, however, but it is not worth the risk.

For the ferret housed outdoors, I would strongly recommend that at least a part of its diet includes raw meat, preferably rabbit meat along with fur and bone if at all possible, though I personally would not recommend giving road casualty victims to your pet, those which are easily located alongside country lanes. You do not know for how long the animal has been dead, for one thing, and, for another, you do not know whether or not the animal was diseased, or, in places where pest control is carried out, whether or not it has eaten poison. I believe that the risks these days are far too great to make this kind of food a viable option. If you do choose to make this kind of food source a part of your pet's diet, especially if you live in a village or a small town surrounded by countryside, then make sure that the animal or bird has been freshly killed, the body may still be warm, or the blood may be quite fresh etc, and remove the innards altogether, just in case of disease.

I believe that dry food is your best option and it is suitable for ferrets kept either inside or outside. Variety though, is essential, and you can change the feed occasionally, or you may wish to give them little treats such as eggs, bread and milk, or specially designed treats. These treats are readily available and are made to be tasty, some containing meat such as chicken, while others contain such things as peanut butter, would you believe! These can be used to supplement their normal diet, or they can be used just as a treat once or twice a day. They are especially useful for giving to your pet directly from your fingers, thus aiding your fitch to accept your handling of it without biting, for feeding it out of your hand is a great way of getting it used to your scent. This will also help them to associate your hands and fingers with pleasureable things, thus they will not resort to biting. Picking your fitch up first and then giving it a treat with the other hand is a great way of getting to know each other. Before giving these treats, however, always read the instructions first, for there is usually a limit to how many you can give per day. Do not be surprised though, if your fitch does not take some of these treats. Personally, I prefer to purchase raw liver and cut this into small pieces and give as a little treat. True, this method is a little messy, but I will guarantee that your fitch will not turn its nose up at this offering, for it is far tastier to them than such things as peanut butter! This kind of treat is also much cheaper than the manufactured kind.

Just another quick word about road casualty victims. *Never* feed rodents such as rats or mice to your pet ferret, even if you find them on a lane right in the heart of the countryside, as these could have eaten poison and this would then be absorbed by your fitch who will probably die, for, by the time you have noticed it is ill, it is already too late and veterinary aid would do little, if anything, to help, except maybe, for putting the poor animal out of its misery by putting it down. At one time I was the caretaker of a country restaurant and any mice or rats which moved in, or even around the building, were swiftly dealt with by the local pest control officer who would regularly come and put poison down all over the place in an attempt to keep rodents out of the restaurant where they would have been a threat to public health. This was out in the depths of the countryside and any road casualties taken from anywhere near this place would be potentially lethal to an unwary animal eating them. In any situation, I would not feed rats to ferrets, for they are disease ridden creatures which carry illnesses that kill man, let alone other species.

Quite a few different animal species are now classed as pests and it is often government policy to deal with them by poisoning, so you must be very careful about feeding any of these pest-listed creatures for this very reason. Of course, in different countries, different animal species will be considered as pests, so this will differ from country to country. True, the feeding of road casualties is totally free of charge, but the risks are not really worth it and, nowadays, with the large variety of other available foods which are quite reasonable in price, this food source is totally unnecessary.

Apart from liver or other forms of raw meat, and manufactured titbits of course, there are many other things which can be used as little treats for your pet fitch. I had a ferret, a dark polecat type, which was crazy about milk chocolate and would not leave me alone whenever I had any. Some fruit may be enjoyed by your fitch, maybe banana or apple or some other kinds. Some may enjoy different varieties of vegetable, or scraps of meat left over at mealtimes, minus any cooked bones for reasons already mentioned.

Whatever you decide to use as little treats for your pets, treats which will undoubtedly help you to forge a closer bond with your ferret and thus guaranteeing a non-biter, make sure that you do not give too many in any one day. This could contribute to your fitch becoming obese, or it could cause digestive problems. So harden your heart and limit the amount of treats you give each day, and, now and then, have a full day when you give no treats at all, just the normal meal, no matter how much your pet pesters you to do otherwise.

When I first began keeping ferrets the free advice given at the time was, that it is best to starve, or fast, your fitch for one day a week, and the argument given in favour of this was very convincing. Living wild, polecats, though they are very efficient and ruthless predators, like other species which prey upon other animals, will not always be successful in their hunting forays and thus may have quite long periods of time when they go without food, hence the reason why countrymen of old always advised having one day a week when you didn't feed your fitch, though fresh water was always available, for water is easily available in the wild, especially in the rain-plagued British Isles!

However, though wild polecats will have days when they fail to catch any prey, don't forget that they 'cache' parts of their kill on other days and so, during leaner times, will be able to return to these food stores and finish off what was left of a previous day's bounty. So it is rather unlikely that wild polecats would go a full day without eating anything at all, not on a regular basis anyway, maybe just the odd occasion when a fox or some other competitor, has found the stored prey and has dug it up and eaten it.

While it may be true that fasting, even for humans, can be beneficial to health (fasting is also said to sharpen the mind), it is unnecessary and will neither harm, nor benefit your fitch, so the choice is entirely up to you. This same argument for fasting one day a week could easily be applied to cats and dogs and, indeed humans, but, think about it, when did you last notice anybody you know starving their cat or dog, or even themselves, for one day a week? Exactly. So you decide whether fasting one day a week, or even one day a fortnight, is something you will apply in the husbandry of your pet ferret. Again, I must emphasize that if you do leave a day when you will not provide food, this will not harm your pet at all. You cannot be accused of cruelty or neglect if you do.

If you decide to feed your young kits on canned food, then I would go for either specially prepared puppy food, or food exclusively for kittens, though you must supplement this with raw meat such as liver cut up into small pieces, raw eggs occasionally, watered down milk once or twice a week, and you may wish to crush up egg shells into a powder and sprinkle this onto the meat, mixing it well in, a few times a week, helping to provide calcium which is good for healthy teeth and bones. Of course, if you know of someone who works their ferrets and who will provide you with a dead rabbit now and then, even better, for freshly killed rabbit will provide all of the goodness your growing kit needs.

Vitamin supplements, if you provide your fitch with a well balanced diet, especially a good quality dry food exclusively for the ferret race, are totally unnecessary and I would not recommend giving them at all, for an overdose of certain vitamins can be harmful, rather than beneficial, to your pet. So concentrate on providing the correct diet, rather than on trying to give them a boost with extra vitamins.

If you feed a quality preparation which is rich in essential oils that are good for the coat of your ferret, making it full of vitality, glossy and shiny, then you need not provide any other supplement which will aid the maintenance of a healthy coat. If you feed canned meat especially, then I would give your fitch just a little cod liver oil, mixed into the food, or just direct from the spoon, as most ferrets love the stuff (yuk!), to help provide the essential ingredients for healthy skin and fur. Cod liver oil is also excellent for aiding the free movement of joints in older ferrets, not to mention humans too.

Unless feeding a dry food which will remain fresh overnight, always remember to remove food from the cage before going to bed, just to avoid any possible stomach upsets, and this is especially important during the warm summer months, for bacteria will multiply rapidly in hot temperatures and this, as anyone with any sense knows, could cause serious harm to a fitch, especially a young or old ferret.

Make sure that your pet receives food regularly, each day. It is far better to put out smaller amounts every day, than to provide a huge feast say every other day. This will ensure that the food is fresh and it will also help your pet to settle into a regular routine, thus it will be much happier in its environment, making it a much better pet to have around the place.

Before moving on to the next chapter, I just want to stress yet again the need to always provide fresh water for your fitch, making sure that you keep a regular check on the amount left in the water bottle, or in the dish you have around the home. And, if your ferret has the run of the place, be certain that it can easily access the room where you have placed the food and the water dish at all times.

A good diet will help keep your pet ferret in tip-top condition, happy and healthy, for years to come, thus the effort you put into making sure that your fitch is getting all that it needs will be generously rewarded as you will enjoy its company, its cheerful, playful nature, throughout its long life, a long life that will give you much pleasure.

Chapter Seven

Housebreaking

Though there can be no doubt that the ferret has been domesticated for thousands of years, mainly as a hunting animal, being used to hunt burrowing creatures in man's quest for survival, there is much uncertainty as to exactly when it was that man first began to exploit the useful qualities of these animals. Some believe that the Egyptians were the first to make use of ferrets, not for the hunting of species suitable as food, but for keeping down rodents around their homes.

Ferrets are most certainly capable of keeping a house free of rats and mice but I find it a little hard to believe that those Egyptians tamed the wild offspring of the Asiatic polecat and had them running around their houses in search of invading rodents. Even if cats had not been domesticated at this time, around 1500BC, dogs certainly were, and it is a well known and well documented fact that small terrier sized dogs were very popular with Egyptian ladies in particular (who would undoubtedly find the smell of ferrets a little unpleasant to say the least, as these, obviously, would not be descented), and these dogs, some of which were used as hunting dogs, would need no help in tackling any rodents found around these homesteads. Anyway, whatever happened in Egypt, in the UK cats became very popular and took over the role of rodent exterminators, except in more remote areas such as on farms where small dogs, alongside the odd cat or two, still patrol the place and get rid of unwelcome pests. Having said this, however, ferrets can squeeze through some incredibly small spaces and so would be far more useful than either dogs or cats for keeping your home free of rodents which can easily destroy a store full of food by leaving droppings, urine and potentially infectious disease all over the place.

So, particularly if you live anywhere near habitat which may have a thriving population of rodents, you may benefit greatly by having a fitch running loose around the place, for it will soon keep them out of your house, though the scent of a ferret around the home is usually enough to keep them out. If you do allow your fitch into the home, either just to spend time with it, give it exercise, or as a permanent arrangement, especially if you live in an apartment and have no choice in the matter, then, if you wish to keep your home clean and relatively free of unpleasant odours, you must housebreak, or house-train, your pet, and this is something that requires diligence and effort on your part, if you are to succeed.

To begin with, it is best to start litter training your ferret inside a sturdy, roomy cage, placing a suitable litter tray, many of which are now specially designed for ferret use and are available from most suppliers, at the farthest end of the cage, opposite the sleeping compartment, for fitches are generally very clean animals and will usually use latrines as far away from their sleeping quarters as possible.

You may wish to use sawdust, or some other type of floor covering in the litter pan, but you must remember that you will need to use the same material when it comes to litter trays being placed around the home a little later on, so sawdust may be a little too messy for you as ferrets will often scratch about in it and kick it all around the tray. If you can put up with this, then sawdust is fine, though specially manufactured ferret litter is now available from some suppliers and you will find this far less of a problem than sawdust which can get all over the place. It is entirely up to you, of course. Some use cat litter, but there has been some concerns about this getting into the eyes of a playful fitch and causing problems which

(*Left*) A litter tray which is perfect for the corner of a room or cage.

(*Right*) Note the high back on this litter tray – this prevents the droppings spilling over.

79

will probably need veterinary care. Though this may be true in a few cases, the majority of ferret owners have had no problems using cat litter and I have had no problems using sawdust either, another material that is loose and flaky and so, in theory, may cause problems.

Ferret litter is possibly the best choice, however, as it is less messy and is designed exclusively for use by fitches. Having placed this in the litter pan which is situated at the far end of the cage, you will now need to be patient and wait until your pet needs to go to the toilet. When about to urinate, or excrete, your fitch will back into a corner with its tail lifted into the air, sometimes rather rapidly, so you must act quickly if it has chosen a spot away from the litter tray. If you are quick enough, get hold of your new pet and place it in the tray instead, though, very often, they will use the correct place without being shown. If you are too late and the fitch has beaten you to it, then do not worry. Simply use a little scoop to pick up the faeces and place it in the litter tray and your ferret should then begin to get the idea, as long as you make sure that you thoroughly clean the area where it made a mistake, so that any lingering odours are removed. Pan training inside the cage is extremely easy when compared to training around the home where the space is much bigger and where there are far more corners to choose from.

Even if your fitch is going to have free run of the house, it is far more effective if you begin litter training in the confined space of a suitable cage. Also, after meals and a few minutes after waking, you may wish to try placing your pet in the litter pan so that it gets the idea of going to it at these times. This will help when it comes to allowing them to have more freedom around the home, though they will rarely want to go to the toilet at times designated by you!

Now that your ferret is regularly using the litter pan inside the cage, it is time to expand on this progress and it is best, once you begin to allow your pet freedom around the home, to restrict its movements to just one room, until it is regularly using the litter tray inside that room, and preferably without mistakes.

After filling the bottom of the litter tray with the material of your choice, place a dropping from the cage on the surface and position the pan in a corner of your choice. Hopefully, the scent of the dropping will attract your pet to the correct corner where it will, if all goes well, use it as a latrine. Again, should your fitch begin to back into the wrong corner, then, if you are quick enough, pick it up and carry it to the tray and, after a while, if you are diligent enough, it should begin to get the idea. If you are not quick enough and your carpet gets soiled, then, after cleaning the area with a little watered-down disinfectant (the type of disinfectant that, when mixed with water, remains clear as there are potentially harmful properties to your fitch, in the type that goes cloudy), you may wish to move the tray to that corner which has been chosen as a latrine by the ferret itself. This may or may not help, for fitches may often choose to use more than one corner when they find themselves in large open spaces around the home.

When accidents do occur, show your displeasure, not by ranting and raving, but by a few stern words, and, equally important, give plenty of praise when things are done correctly, maybe even providing a little treat every time the litter pan is used. Ferrets are intelligent creatures and will eventually learn to associate praise and rewards with using the pan and unpleasant, disapproving words whenever they go wrong and mess outside the litter tray. Though, of course, this may take a while to accomplish and, don't forget, some will learn much sooner than others, so do not be discouraged if one of your fitches takes an age to catch on.

Now that your ferret is regularly using the litter tray and accidents are very few, or even non-existent, it is time to open up another room for your pet to enjoy, though the number of rooms it will be allowed to play in is entirely up to you, but you must make sure that there is at least one litter tray placed in a convenient corner of each of these rooms, for it is unlikely that the fitch will travel all the way back to the other room just to use the toilet, especially if it is in a room at the other end of the home. And some rooms, especially the larger ones, may need to have two litter trays placed in two separate corners, preferably at opposite ends of the room. If you fail to provide adequate toilet facilities, then your efforts will probably turn out to be in vain, for a ferret will soon get out of the habit of using a pan when there are very few around, but there are plenty of other, more convenient corners which can be used instead, so it is very important to get this part of the training programme right, and you must get it right the first time. Inconsistency will only confuse the poor animal.

When it is time to begin allowing your fitch to use rooms around the home, it is very, very important that you make sure that each of these rooms are what is known as 'ferret-proof'. That is, you must make absolutely certain that every room where access is allowed is a safe environment for your fitch to play in. Check, first of all, for holes in the wall, or around the floor area, and fill these in. In the kitchen especially, make sure that the holes allowing pipes to go through the wall, or down under the floor, are only big enough for the pipes themselves. Any all-purpose filler will do the job perfectly well and the time and effort involved will be well worth it, for it may mean the difference between a safe and sound pet, or an escapee which you will probably never see again and which will more than likely be run over on a busy road, or may simply starve to death, or even die of the cold. So, again, be very diligent about this matter and, if the thought of crawling about on your hands and knees searching for holes isn't very appealing, then either get someone to do it for you (but beware, when their pet is not involved, other people will not be as diligent in finding holes around the place), or do not allow your fitch the freedom of your home. Do not forget that ferrets have been used by man for centuries to seek out prey from their lairs inside tunnels deep underground, and flushing it out into the open. Indeed, an early description of the use of terriers in the same manner reads thus: 'Another sorte there is which hunteth the fox and the badger or greye onely, whom we call terrars, because they (after the manner and custome of FERRETS in searching for counyes) creep into the ground, and by that means make afrayde, nyppe and bite the foxe and the badger . . .' So ferrets are naturally drawn to dark passages, as is the terrier, and will then go on to exploring these passages, for they are very curious creatures and are like the little old lady who gads about town in search of the lastest gossip. If you are to enjoy peace of mind when your pet fitch is enjoying the run of your home, then make sure that all of these holes are filled in and always be aware that, especially where pipes go through the wall, the fitch may scratch and chew around these in order to make a bigger hole for itself. If you notice your pet spending a lot of time out of sight behind a cupboard, or some other item of furniture, then it may be well worth taking a peep to see what it is up to.

When ferrets are running free around the home, it is always best to keep windows either closed, or, where there are windows which will easily fix into position and cannot thereafter be moved to open further, left slightly ajar, for ferrets are surprisingly good climbers and an open window will often be within easy reach. There may also be some concerns about cats getting in through an open window and attacking your fitch, though this is most unlikely. The countryman out ferreting for rabbits will often bolt cats, usually feral cats (domestic cats turned wild) which are of a fiercer nature than a house tabby. So it is not likely that a cat will attack a fitch. However, a kit would be far more at risk, so it is well worth taking these precautions anyway, just to be sure and to give you peace of mind, even though cats will not usually tackle anything bigger than a rat and I have seen squirrels seeing the local cat population off around my home.

Having filled in all of the holes around your home and having checked and filled in any large gaps around the pipes, the windows having been secured so that escape, even for a fitch which can squeeze through some incredibly tight spaces, is an impossibility, the final thing to do is to make sure that any wires for your electrical appliances are not in places where your fitch can be out of sight having a good old chew and killing itself in the process. Where these wires do trail behind cupboards or any other items of furniture, it is a good idea to place hooks on the backs of these so that the wires can be hung on these, thus keeping them off the ground and out of reach of your pet. I have to say that it is most unlikely that a ferret will take to chewing wires, but some do develop bad habits and the odd one may take to this lethal pastime, so always be aware of your fitch spending time out of your sight, for it may be up to no good.

It is also well worth checking electrical appliances such as fridges and washing machines as your fitch may be able to get inside these and, obviously, in so doing is putting itself in great danger of electrocuting itself and this, of course, could also start a fire in your home. So, unless you can find an effective way of stopping your pet from getting behind such appliances it is probably best not to allow your ferret the run of the kitchen or the laundry room. It may also be well worth turning off any electrical appliances at the socket, or even unplugging them when you are out if you allow your pet the run of the place and do not cage at all.

You must be as safety conscious whenever your fitch is free around the house, as you would be with a young child, or a puppy or kitten, and this also means keeping such things as buckets empty of water when out of use as a ferret may fall into such things and drown.

If you have trouble with litter pans being knocked over by a mischievous fitch, then you may need to fix them down. This can be done with wire inside the cage, or with doublesided tape around the house.

Chapter Eight

Exercising Your Ferret

As long as a cage is roomy enough, exercise begins within the confines of a ferret's living quarters, though that cannot possibly be the be-all and end-all of the matter, for ferrets are highly active creatures with a measure of intelligence and they will need to be stimulated both physically and mentally, if they are to remain at their happiest and in the best of condition. For instance, a polecat living in the wild will cover great distances in its search for prey and, in the spring months, for a mate, using both its physical energies and its mental faculties in so doing. A ferret, don't forget, is simply a domesticated polecat and will need physical and mental exercise just the same, so providing this exercise is one of the more important aspects of ferret keeping and their care.

Roomy cages are essential.

Providing a roomy enough cage is absolutely vital, but, on its own, is not enough. You must also provide other alternatives and the first place you can look to provide this extra space is, of course, your home. You may be one of those who allows your pet the free run of the home at all times and in this sort of situation a ferret will enjoy the wider spaces available, though many choose to house their ferrets inside a cage while still keeping them indoors, allowing them to enjoy free run of the place at more convenient times. Others, such as myself, house them outside, but undercover, inside a shed or an outhouse, or they cage them out in the yard or garden. For those who cage their pets in a shed, a good way of providing extra room for exercise is to allow them the run of the shed or outhouse, either at convenient times throughout the day, or even permanently, with a run leading from the cage to the floor for easy access. They will then have both their cages and the shed to play in and this is the basic requirement for ferret exercise.

Whatever situation you choose for housing ferrets, allowing them into your home, either permanently, or just for providing extra exercise for the fitch, is essential really, for this allows you to spend much time with your pet who will need a great deal of human contact if it is to be properly socialized and thus trustworthy around humans, especially with children. Children will play with ferrets for hours around the home and this is good stimulation for both parties involved.

Some owners do not allow a ferret into the home, especially a fully grown male which has not been descented, but, if you do have your fitch indoors with the family, then I cannot stress enough the need to ferret-proof those rooms where it will be running free. Always be aware of both the safety of your pet, and any opportunities which may arise when it could escape, especially when children are around, for they will often leave doors open and a ferret, whose very nature makes them very agile and nimble and thus expert escape artists, could easily slip away unseen and be lost forever. Make no mistake, if the opportunity arises, ferrets can, and do, escape.

A friend of mine, Susan Nuttall, has just told me an amusing and interesting tale about an escapee ferret which caused havoc among parishioners who were gathered for worship one Sunday morning.

While the service was in full swing things suddenly became rather lively when one of the girls sitting next to Susan jumped up screaming, her arms shooting up into the air and knocking Susan's glasses off, and there, wandering around the floor, was a ferret which had obviously escaped. Though it wasn't amusing at the time, Susan laughed as she told me of how she could tell where the ferret was by the worshippers, one after the other, jumping out of their seats as the fitch made its way around, and, at one point, climbed the steps and attempted to get up the vestment of the clergyman. Eventually, after much fuss, they managed to shepherd it outside and it was last seen wandering down the road.

Anyone who has had any contact with ferrets and knows their natures would undoubtedly have simply picked up the animal and taken it to the nearest animal shelter. However, most people fear these animals, as was illustrated in this story, and will not go near them, their fear no doubt fuelled by the image portrayed by the media and television. So be extra vigilant about making sure your ferret's environment is both safe and secure.

For those who will not have a ferret in the home then the garden, or yard, is a good place to exercise your pet, though, again, you must be very careful about securing the exercise area, never leaving your fitch unattended at any time. Safety awareness is essential and if you have a pond in the garden, then make sure that, should your pet decide to go in for a swim, or should it fall in, that it can easily get out again. This should already have been done however, for creatures such as frogs, lizards and hedgehogs need to be able to both access and exit a pond, for hedgehogs will sometimes fall into these as they are rather clumsy animals, and so must be able to get out if they are to avoid drowning. Placing a few stones in a sort of step formation by the edge of the water is an ideal way of making sure your pond is not a deathtrap to either wildlife or your new pet. When exercising a ferret around the garden, especially when a young kit, be aware also of cats lurking in the borders with designs on your fitch, though trouble from cats is rare and few will even go near a ferret, let alone attempt to tackle one.

Exercising your fitch in the garden, however, can be a problem during prolonged periods of bad weather and especially during the winter when you will not be too keen on spending any amount of time outside in the cold, so this arrangement is rather limited and, rather than not allowing your pet into the home at all, maybe you could exercise it outside whenever possible, and then bring it indoors when the weather is too unpleasant to be out of doors. During the summer months, of course, it will be a pleasure spending time outside with your pet and no doubt you can get on with a few jobs around the garden at the same time, as long as you keep a close eye on the whereabouts of the ferret at all times.

Having sorted out the arrangements for the basic requirement of your ferret's needs in the exercise department, providing plenty of space for the animal to run around and thus burn off the excess energy, of which there will be no shortage I can assure you, it is now time to make sure that there is plenty of mental stimulation too. Polecats out in the countryside, whatever country they inhabit, will find themselves searching out every nook and cranny in their bid to stay alive, often against great odds, and exploring numerous tunnels in search of prey such as rabbits. When you watch your fitch as it runs around your home, or outside in the garden, you will soon notice that it behaves in the same manner, curiously seeking out games to play, or bits of food, at every little dark place it comes to. This is only natural and, though ferrets have been domesticated for millenniums, this behaviour betrays their wild background and that is why man has been able to use this animal for both pest control and for providing fresh meat for the family. It is this natural instinct that you must cater for by providing, not only an exercise area of suitable size, but also obstacles and maybe even toys which will help to stimulate the mental faculties and thus keep your pet happy.

If the cage is roomy enough, then you may wish to place obstacles such as small plastic pipes (the cardboard centres of toilet rolls and kitchen paper rolls are also ideal) inside for when you are not around. This is quite important if you are out all day and you are only able to spend time in the evenings with your pet, but is less of a necessity if you are at home for quite long periods and give your ferret plenty of attention. Whatever you do, it is essential to provide these in the main exercise area, if not in the actual cage. Plastic drain-pipes which are large enough for your fitch to get through, are ideal toys. If you join a few pipes together and make a tunnel system, placing little treats inside on occasions, this will be great for keeping ferrets occupied for long periods of time. These pipes are very cheap to buy new, but a builders' reclamation yard may prove much cheaper and you will be able to get these for next to nothing.

Of course, you could make your own tunnel system, either out in the garden or inside your home, by using materials found around the home, and this is the system I used when I was a lad. I was an avid collector of books and I would pile them up in long rows on the bedroom floor with a gap between them of around two or three inches and then, when the books were piled high enough, I would place heavier books across the top so that the space betwixt was now enclosed and so could be used as a tunnel system for my ferrets. I placed little treats in well chosen locations inside these tunnels and then in would go a fitch, searching every passageway for the treats, in exactly the same manner as the wild polecat searching out its eagerly sought prey. Most of my ferrets loved to run through these tunnels and dark passageways which I frequently changed around to stop them getting bored with the game, but Ben, though he loved to explore these dark places as any self respecting fitch does, was much too large for these and would often demolish the tunnels as he went through them, squeezing his bulk along with a book very often on his back. I was forever repairing this tunnel system whenever Ben was around.

Though this system worked for me and provided my ferrets with

If you cannot make your garden escape proof, this wire enclosure is ideal for exercising ferrets outdoors, though you must provide shade from the sun and fresh water, and a warm nesting box in colder weather.

endless fun and activity, stimulating them both mentally and physically, it is a lot of messing putting books back on the shelves and
then pulling them off again the next day, so the plastic drainpipes
are far superior really. It is well worth investing in a few lengths of
these pipes, including bends, joints and junctions, so that you can
make up a tunnel system and change it around at times for variety,
rather than just having one or two straight lengths for them to run
through. Especially if you live in a small apartment, these pipes are
most useful, for they can easily be stored away and will take up very
little space.

Other toys can be obtained via suppliers which you will find listed
in the back of this book, though their uses are rather limited as
ferrets, just like puppies and kittens, will soon get bored with these
and very often will find your furniture, your curtains and other
household items of much more interest, though you may wish to
select one or two toys and see how your fitch responds to them. If
your ferret has to spend long periods of time during the day inside
its cage, these toys may help to provide a little activity that will help
allay boredom, though it is far better to have two living together, so
that they can play while you are out for the day. This is not in any
way a hard and fast rule for I have found that ferrets caged alone
will do just as well as those caged together, as long as you spend time
with them and provide plenty of exercise. Even if you keep more
than one ferret, you may wish to cage them separately, but bring
them together at exercise time, for they are sociable creatures and
will therefore enjoy the company of one another, playing
and cavorting for hours at a time.

The cage area, the home, even if you only allow them the run of
one room, or the garden if it is secure enough for holding a fitch, are
the basic requirements for the physical and mental stimulation that
will aid your new pet to live a happy and content life, but there are
more places you can use if you so wish. Even big cities have public
parks, green, wide open spaces where you may wish to take your
fitch for additional exercise, or maybe just a change of scenery, an
escape from the usual routine. Before venturing out to your local
park however, or anywhere else out of doors for that matter, you
need to purchase a harness and a lead in order to prevent your fitch
from wandering off. Don't forget, ferrets are wick* animals and

* agile, quick, alert

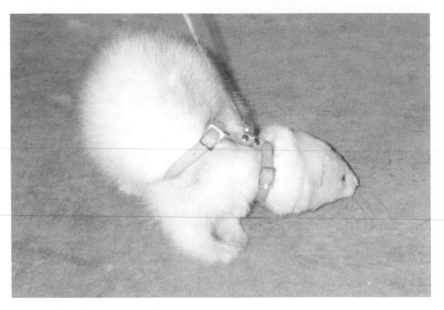

A harness is essential when exercising away from home.

could easily get into undergrowth before you can prevent it, so you must secure your ferret in this way, rather than allowing it to run free as you would in the garden. A collar will, it is true, do the same job as a harness, but it will not be as secure, for collars can easily slip off, or a ferret, after a determined effort, may slip out of it. No matter how agile a fitch is, it will find it impossible to wriggle free of the confines of a properly secured harness and so this is the best method to use when exercising your pet outdoors. Both collars and harnesses are readily available from most pet shops and suppliers.

You may live near a good stretch of countryside and no doubt you will enjoy taking your ferret to such places and walking it there, allowing it to explore the hidden delights to be found there. Wherever you choose to go in order to provide that extra exercise, always be aware that, whenever you are in public places, it is very likely that there will be dogs around also, so you must be alert at all times for some dogs, particularly terriers which still have a strong hunting instinct in many breeds, will readily attack a fitch and kill it, or, at best, severely injure it. This is another good reason for

90

fitting a harness rather than a collar because if you notice a dog preparing for attack, or approaching, you can quickly respond by lifting your ferret out of harm's way with the lead itself, saving much valuable time having to bend down in order to get hold of your pet, though you must be careful not to panic and jerk your ferret all over the place. Obviously, this should not be done when a fitch is wearing a collar as you may do more harm than good, so go for the harness which is the better option and, though it is more complicated to fit than a collar, you will soon learn how to fit a harness in no time at all.

If you find it impossible to take your pet ferret out into these open spaces for exercise and maybe you live in a small apartment with no yard or garden, then do not despair, for a ferret will be happy enough as long as you spend time with it and allow it out of its cage for a run around the home. If you play with it and provide interesting things for it to do, such as the tunnel system mentioned earlier, then it will be content with its lot in life, though, in limited spaces, especially if you are out at school, or work, all day, as I have already said, you may wish to keep two ferrets. They will burn off more energy as they play together when out at exercise around your home. Again, this is not to be taken as a hard and fast rule, it is just something worth thinking about.

Because of their sociable, friendly nature, ferrets around the home are great company and will cheer the place up greatly, something that will be very welcome for those who live alone and get a little lonely at times, and they will provide you with endless amounts of pleasure and amusement as they bound around playing their silly games, either with you, or, where two or more are kept, with each other. For younger ones, the keeping of ferrets will not only provide you with much fun, but will also help teach responsibility, for a pet fitch will depend on its owner for food, drink, a clean and safe environment, and love and attention. I once watched a young lad who was out exercising a large hob ferret and I could see clearly in his face the pleasure he obtained from watching his pet as it explored every little hole along a wall, using its nose in search of any titbits of food that may be found lurking in these little hidden places. Pets are good for children and pet ferrets will be a delight to any youngster who will probably lavish more time and attention on them than most adults, making sure they have the care and the exercise that is necessary for a fitch to live life to the full.

If you do take your ferrets out of the home environment, then you must make sure that you transport them safely. It is the transportation of ferrets that is the subject of the next chapter. However, before we discuss this important aspect of ferret husbandry, I just want to mention the importance of not trespassing in places you choose as exercise areas for your ferret. Always check that public access is allowed before using a place for this purpose and, if necessary, ask the landowner for permission to use their land, explaining that your ferret will be secured to a lead and harness for many people, especially farmers, would be very reluctant to allow you onto their land in order to exercise your fitch if there was any chance of it escaping and getting in among their chickens or ducks. Having permission in this way ensures that you do not get into trouble for trespassing on someone's land.

Chapter Nine

Transporting Your Ferret

Many ferret owners enjoy taking their pets to places that are suitable for extra exercise, but getting there means that you will need to transport your fitch and you must make sure that it is both safe and comfortable when doing so.

Of course, the larger hob ferret will need a bigger space than the smaller jill, and there is more than one option available for the purpose of transporting these to and from your favourite exercise area. You could quite easily construct your own carrying box and this would indeed keep the costs down, but there are some excellent ready made boxes exclusively for transporting ferrets and these are ideal, though they can be a little bulky and are not really suitable for carrying around town and they are especially awkward for getting on and off public transport, or when making your way through a

A well constructed wooden carrying box with bedding, fresh water and ventilation.

bustling crowd of people. However, for use in the car, they are ideal and can be purchased from any reputable supplier, though some can be a little on the expensive side.

Wooden boxes are the best for this purpose, especially those made from lightweight wood, as these are strong enough to prevent a ferret from making an escape route, and they have the added attraction of being strong enough to protect your fitch should you have a fall while out carrying the animal, avoiding crushing it to death or causing serious injury. As I have said though, these can be very expensive and so you may wish to look at other options. For instance, on the British market at the time of writing, a ferret carrying box designed for two and made out of lightweight wood,

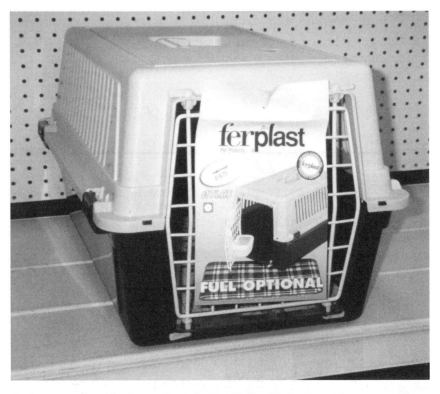

A plastic carrier with plenty of ventilation. Enclosed plastic carriers are not suitable as these produce much condensation and do not keep cool in warm weather.

will set you back around £25 to £30, and possibly more from some suppliers. These are quite bulky and are most certainly not suitable for transporting ferrets across town, for you will surely injure somebody with the hard corners of the box when making your way down a busy street. These wooden boxes are for the ferreting man who tramps the countryside in search of rats or rabbits, and are not really for the pet ferret keeper.

Specially designed plastic pet carriers are fine but plastic boxes are not really suitable, especially in hot weather as, even if you put plenty of holes in the sides of the box for ventilation, the body heat of the ferret will produce much condensation inside these and, anyway, may shatter if you were to drop it or have a fall. I did use a plastic box for carrying my ferrets at one time, but only for a very brief spell. Despite the several ventilation holes I put in the sides of this box, I could not stop the problem of condensation building up inside and I did have problems with the catch coming undone, though it was very lightweight and far less bulky than a wooden box,

A wooden carrying box with plenty of ventilation. Note the shredded paper inside for bedding.

95

but, unfortunately, the advantages do not outweigh the disadvantages and so I soon stopped using plastic.

There are some very good cardboard carrying boxes available from pet shops and stores, but these are more for use when visiting the vet for pets such as cats or rabbits, rather than as carrying boxes for fitches. Ferrets will often soil when in transit and a cardboard carrier will soon become ruined when in regular use. Also, if you were to fall the cardboard would give very limited protection to your pet.

You could, of course, use a sack made of suitably strong material, the type with strings which pull the top tight together in the same manner as a string purse, but the trouble with this method is that it obviously gives no protection to a fitch should you fall, or while jostling through a large crowd of people milling around on a busy street, and your fitch, if it soils, will have to lie in it. Ferrets are very clean animals and will not appreciate having to lie in their excrement and urine, so a sack with purse strings isn't really suitable. This method of carrying a ferret was favoured by poachers who could conceal the sack and its contents if necessary, and because sacks are very lightweight and not bulky. Also, if the gamekeeper chanced upon the poacher while trespassing on the squire's land, the sack

A bow-back wooden carrying box, for extra comfort for the carrier.

Another style of wooden carrier. (A large hob ferret sleeps inside)

would not hinder a speedy exit. Many a time a poacher would be disturbed by the gamekeeper doing his rounds while the fitch was below ground hunting rabbits, and these would then be left in the burrow by the owner who would rather find himself with one less ferret, than end up in the dock of the local magistrates' court. These ferrets would easily survive, living off rabbits and birds, and this undoubtedly helped the recolonization of the wild polecat in the British countryside. One can easily imagine that escape for a poacher caught in the act would be nigh-on impossible if he had to carry a large, bulky wooden box around the countryside with him.

A medium sized travel bag makes an ideal carrier about town and your fitch, if you drive part of the way, can easily be transferred from the wooden box in the car, into this, though, in large crowds, these bags can be rather flimsy and give very little precious protection in the crush of a crowd of people, especially while on the Underground which can be very crowded indeed. A sudden jolt from the train could cause your fitch to be crushed in this sort of carrier, so I suggest that you make a little wooden frame to go inside the travel bag and thus strengthen the sides. This takes very little expense, time or effort and is well worth the peace of mind you will enjoy while out and about, for, should the worst happen, your ferrets will be secured safely inside with a protective frame around them, much the same as a racing car with roll bars protects the driver should he crash. You needn't worry about hard edges hurting people either, for the frame fits inside the bag and so the edges of the wood are cushioned by the material of the bag itself.

You could easily make this frame yourself, though it will cost very little to have a professional do it for you. Provide the carpenter with both the bag and the timber, so that he can get the measurements just right. You should use one inch by one inch timber and you will need to cut six pieces for the length of the inside of the travel bag, having a pair of lengths at the top, the bottom and in the middle for extra strength, and you will need to cut nine pieces for the width, with three at each end, top, bottom and middle, and three in the middle of the bag, to prevent it from being crushed should the worst happen. You should fix the whole frame together either with nails and a good quality glue construction or with screws and glue. Whether you use nails or screws, make absolutely certain that your fitch is safe by checking that none are sticking out of the wood in any place whatsoever. If the job is done by a professional, this is most unlikely, though still check anyway, but there may be one or two sticking out of the wood if you choose to do the job yourself, for, if you are anything like me, you may knock one or two nails in at the wrong angle and the end may protrude from the frame at some

A superb four unit carrier ideal for the back of a car and for use at shows.

A plastic carrier. Make sure the mesh on the door does not allow a ferret to escape. The albino seems to be asking to be let out!

point, so check, and then double check. You will also need two pieces of wood at each end for height.

Whatever you decide upon for transporting your fitch, make sure that it is roomy enough, it is escape proof, is properly ventilated and is strong and sturdy enough to cope in a crushing crowd. Also, when you travel by car, please do not leave your ferret in your car in hot weather. Always remember that hot cars kill animals very quickly indeed.

Chapter Ten

Hygiene and Health

Ferrets are full of life and vitality and the way they bound around the home, ready for play at almost any time, demonstrates this and betrays their happy, friendly nature. True, not all ferrets are very playful, for they have different personalities as anyone who has kept them in large numbers will know, but all do have one common link and that is that, for a fitch to be at its happiest, it must be healthy, whatever its personality, and to keep it healthy requires diligence on your part. This principal is true of any kind of creature so you must do your utmost to keep your ferret in good condition and the basic requirements for doing so are: a proper diet which includes readily available fresh water, the correct amount of exercise and mental stimulation, and, of course, a clean environment where diseases cannot breed.

Cleanliness begins at home and the cage is the first place to look when it comes to good ferret husbandry. Cleaning the cage is so simple and quick to do that it is a crime to be neglectful in this department. The litter tray, placed in a corner, or right across the end of the cage that is the furthest away from the sleeping quarters, is very easily and quickly cleaned. If you have a yard, garden, or a balcony outside an apartment, it is a good idea to have a small bin lined with a plastic bin bag just outside the home which you can use to hold the waste from the cage and this should be emptied from the litter tray, whether it be in the cage, or around the home, at least once a day, though a minimum of twice a day, particularly if you keep more than one ferret, would be far better. I use a wallpaper scraper for cleaning out and this can be used for both the litter pan and the cage floor too.

If you have a mixture of disinfectant and water inside a spray bottle, the kind that is readily available at any store which sells household cleaners, you can quickly spray the pan and give it a wipe over with an old cloth, and thus remove very swiftly any lingering

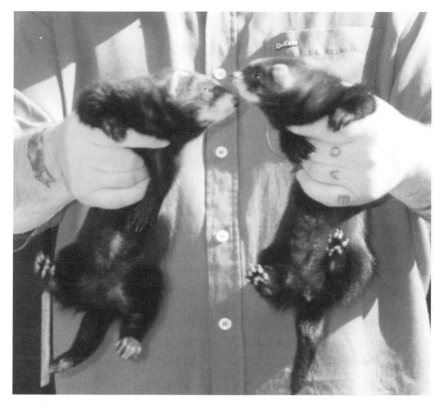

Young hob and jill polecats.

odours which may lurk around the home. Removing and cleaning the litter pan once or twice each day will not only help to keep your fitch in good health, but will also help to keep your home free of ferret odours – something which must appeal to any who have even the slightest bit of pride in their home.

The actual cage itself will need to be cleaned out at least once a week, provided your fitch has learnt to use the litter pan, rather than the floor of the cage, as a latrine. Of course, you may find, particularly around the home if you cage your pet indoors with you, that odours are a little too strong when cleaning just once a week, so up the cleaning sessions to twice, or even three times a week. Again, the more you clean, the more effective you will be in tackling the problems of odours around your home. However, if you cage outdoors,

101

or inside a shed, or outhouse, then it shouldn't be necessary to clean the cage more than once a week, though you must be just as diligent with cleaning out the litter pan as a build-up of droppings will do nothing for the health and general well-being of your pet.

Using a wallpaper scraper, or something similar, scrape the sawdust, or whatever material you have chosen to use, onto a small coal shovel and tip into your bin which you use exclusively for this purpose (if you do not have anywhere outside for such a bin, then you must find a convenient corner indoors, though you must seal the bag again after each use in order to prevent bad smells from invading your apartment), and then, once you have thoroughly cleaned it out, spray the floor and walls of the cage only lightly, not soaking them, and leave to dry for a while, exercising your fitch at the same time, until the cage is ready for it again. It won't take long to dry, provided you do not wet it too much, though I would give the cage a thorough soaking with the water and disinfectant once a month.

If you use a cage made entirely of wire or mesh then it is still a good idea to give it a spray and a wipe over with an old cloth, at least once a week. This type of cage is the best type for keeping clean, but I have had no problems whatsoever using wooden cages and I have kept many, many ferrets in this type of cage.

Kits at nine weeks of age.

Another thing to be aware of when it comes to doing your bit to keep a ferret happy and healthy is to check for any bits of food that have been stored away. Look especially inside the sleeping box – a favourite place for a fitch to 'cache' its food, and under any piles of sawdust, or other materials used, particularly in corners. Once found, remove and throw away these stale leftovers from mealtime, unless, of course, it is a dry food which is safe to leave for upwards of a couple of days, though I would still remove it and place it back inside the food dish.

Do not forget that ferrets are very clean animals which come from a very clean family, so you must think like them. A ferret will usually hate to be anywhere near its droppings for instance, so allowing them to pile up for days on end so that your pet is almost sleeping next to them, is not good at all, in fact, it is downright neglectful and cruel, so make certain that the environment your ferret lives in and exercises in is both safe and clean. This is also good for you and your family, for odours will undoubtedly be kept down to a minimum and you may even find that, if you are diligent in this regard, which I am sure you will be, you may not need to have your fitch descented. True the male fitch will smell quite strongly once it reaches sexual maturity and so, quite frankly, you may have little choice but to have it descented by having it castrated as well as by having the anal glands removed, though you may find, as many others have, that you can put up with a certain amount of ferret odour, so descenting may not be an option you will go for. Jill ferrets are far less smelly than males and many can put up with their odour, as long as they take certain measures to reduce that odour some-what, measures which can be most effective if, again, you are diligent.

Bathing a ferret will without doubt aid in the fight against odours permeating your home, but, it has to be said, it won't take too long for your fitch to be back to its old self. You could, of course, bathe your ferret on a regular basis, but this is not a good idea, especially with those housed outside, for bathing reduces the natural oils in the coat and these oils help protect against the cold. It is not really advisable to bathe too frequently in case the animal catches a bad cold which could lead to its death.

Many use shampoos manufactured for cats, but I have found that plain old washing up liquid will do the job just as effectively. This will also kill fleas, but it will not get rid of the eggs (more about flea

103

infestation a little later on). It will also aid the reduction of the odour problem on a temporary basis, so other measures are necessary.

Using an odour neutralizer on your carpet in the rooms where your fitch runs free is another way of tackling unwanted odours, but you must make sure that you vacuum it up thoroughly after use, to avoid any harm which may occur, however unlikely, to your pet. Also, you can burn scented candles, or have dishes, well out of reach of a mischievous fitch of course, of pot pourri scented with natural oils which are easily obtained and are very reasonably priced too. I avoid chemical air fresheners generally, and go for natural ways of freshening the air in my home. If you have room to grow them, sweet peas are an excellent way of freshening your home naturally. Just cut a few of the sweet smelling flowers on a regular basis throughout the growing season and place them in a vase with clean water. The more flowers you cut, the more flowers will appear on your plant, and they really are effective as natural air fresheners.

If, however, after having tried all of these methods, you still find odours lingering and you cannot put up with the situation, then it is not too late to have your pet descented. Ferret odours can linger on your clothing after you have been handling them and many find that they just cannot put up with this. Descenting may be the only option available to you if you find the smell intolerable.

Bathing

Bathing a ferret is both simple and a pleasure. Make certain that the water is slightly warm to the touch, not hot. You can bathe it in the sink, the bath, or even in the well of the shower. I prefer to use the bath and you should have the water deep enough so that your pet can both enjoy a swim, and touch the floor of the bath. A bath mat will help it to get a grip as it walks in the water. Using a little washing up liquid, gently work it into the fur all over the body and up the neck, but without getting any around the head, especially the eyes and ears, until you work up a lather, and then rinse off with clean water, again avoiding getting any into the eyes or the ears. Now it is time to dry off your fitch. If you want to allow it to spend a bit of time swimming in the water, then do so *before* you use either washing up liquid, or shampoo, so as to avoid it getting soap into its eyes. Once it has been washed and rinsed, then remove it from the water *immediately*.

Using a towel specifically put aside for your pet, gently rub it until the fur feels just damp and then allow it the freedom of a room which has already been heated up *before* you put the fitch into it, and it will run around and play games and lick itself until it is at last dry. *Never* put a ferret back in its cage outside the home while it is still wet, for it could easily chill and die if you do so. It will take around two hours for it to dry properly, so allow for this time *before* you decide to bathe. It is no good bathing a fitch at eleven thirty if you are going out at twelve, though, of course, you may be tempted to use methods which can easily dry a fitch in no time at all, such as a hairdryer.

I would not recommend the use of a hairdryer, or any other method of quick drying for that matter, and for one very good reason. Hairdryers get very hot and, even though you can dry a ferret from a distance, it cannot tell you if it is either too warm, or too cold. It is far better to give it a good, gentle rub down, and then allow it to dry naturally inside a warmed room. A room does not need to be heated so that it feels like a tropical paradise, but it must be heated enough to take the chill out of it.

Just another quick word about bathing. *Never, never, never* leave your ferret unattended, even for a minute, while it is being bathed. You must think as you would with a small child and be safety conscious at all times.

Mites, Fleas, Worms

For cleaning a ferret's ears, you will need to use either cotton wool, or, even better, cotton buds, but only clean the flaps of the ears, including behind them where muck can accumulate, and *never* attempt to clean inside the ear, for you could go poking around and do irreparable damage. If your fitch constantly scratches its ears, or you notice they are very dirty, perhaps even bleeding a little with the constant scratching, then consult a veterinary surgeon every time. The same goes for the eyes. If there are any problems with these, apart from getting a little mucky on occasion, something which can be dealt with very easily using cotton wool dipped in clean tepid water and squeezed out until damp, then allow the vet to deal with it. Eyes and ears are very delicate and are best left to the professionals. Mites can sometimes invade the ears of a fitch and, again, the vet will provide drops to clear them up.

Fleas can breed until they reach plague proportions and, though they will not live on humans, can and do bite them, and there is nothing worse than sitting eating your food, or drinking a cup of tea, and having a flea land in it while it is jumping around the place! You need to take action and preventative steps are best to sort this problem quickly, for flea infestation is very uncomfortable for a fitch as it constantly scratches and nibbles in order to attempt to get rid of the irritation. This scratching can then lead to sores on the skin which can become infected, so it is best to act quickly. Again, many use remedies designed for cats, but there are always new products emerging on the market and you may, if you shop around, find something suitable for ferrets. Before you do, seek veterinary advice for your vet will stock a wide range of remedies which are far more effective than those available at pet stores, and the vet will also be able to advise on the safest products and the safest doses to use. Veterinary treatment for ferrets is, in fact, inexpensive, so professional advice is always the best and it is better to be safe than sorry.

The same advice goes for worming your pet ferret. True, you could use remedies for cats, but you will have to guess the correct dosage, so seek veterinary treatment which will, again, be far more effective then shop-bought treatments. If your ferret has worms, you may see little white segments in its droppings, or it may look underweight, the shine gone out of its coat. Again, do not seek pet shop treatments, go to your vet and make sure your pet gets the best, and the correct, treatment available. The cost will not be much, but it will be money well spent. Even if you come across products designed specifically for ferrets, still seek veterinary advice first, for, as I have said, the most effective, and the safest, treatments, will always be found at your local veterinary centre.

Fleas and worms will not usually be too troublesome, though trouble with worms may follow a flea infestation, so, if you tackle the flea problem first, preventing, rather than curing, you will have less trouble with worms in your ferrets as a result. Fleas can live on carpets and furniture so ask your vet about ways to combat this problem. One of the things you can do in order to drastically reduce the flea problem is to keep a cat flea collar in the bag of your vacuum cleaner which will kill those fleas that are picked up while you are cleaning. This is best done once a day as it will deal with newly hatched fleas before they become too much of a problem. These flea collars are quite inexpensive, though you must renew them when their lifespan runs out.

This may seem like a lot of advice to seek from your vet and you may be a little alarmed, thinking of all those visits and the cost of such, but all of this information can be gleaned from just one visit and, if you ask about these things while you visit your vet to have your fitch vaccinated, then you may find that you get that advice free of charge.

Vaccination

Vaccination against distemper is necessary in those countries where it is readily available for use on ferrets. Distemper is most commonly caught from being in contact with infected dogs, though they are also susceptible to feline distemper. Even if you do not exercise your fitch out of doors, it may still be the best policy to vaccinate, though the need for yearly boosters is, I have to say, rather controversial. I personally do not have yearly boosters for my animals as many within the profession believe that the antibodies from a single vaccination will last a lifetime. I have had no problems in over two decades by getting my animals vaccinated just once in their lifetime, though I am not telling you to do the same. You may opt for yearly boosters, for peace of mind if nothing else, and that is entirely your decision. The important thing is to get vaccinations for your ferret where they are available. Distemper in ferrets is usually always fatal, so, if your fitch does contract this disease before you have a chance to get it vaccinated, then the kindest thing to do is to have it put to sleep. This is totally painless and a very humane way of putting an animal out of its suffering. For feline distemper a ferret should be innoculated with killed tissue vaccine, though the vet should be well aware of this.

The symptoms that will appear if your fitch contracts distemper are as follows: runny nose and eyes, going off food, diarrhoea, drinking a lot and vomiting, possibly with convulsions, though this usually occurs in the latter stages. The eyes will usually be full of matter too. If you suspect this disease in your pet, then isolate it from other animals and get it to the vet as quickly as possible. And remember to wash your hands with a disinfectant after handling the infected animal. The cage and nesting box will also need to be thoroughly disinfected and all the materials from the cage should be burned if possible. If this is not an option for you, then spray thoroughly with a strong disinfectant and seal it in a plastic binliner.

You must do your utmost to avoid spreading this highly infectious disease.

If you live in a country where rabies is considered a problem, then it may be best to have your fitch vaccinated against this disease also. Again, killed tissue vaccine should be used, otherwise the vaccine may actually cause rabies in a ferret. Of course, your vet may consider it unnecessary to vaccinate, so seek his advice on this matter. A lack of energy, anxiety and posterior paralysis are the main symptoms to look for, but the chances of a fitch, especially one that only exercises in or around the home, contracting this disease are practically nil.

Colds and 'Flu

Ferrets, like humans, can catch common colds and 'flu, and this can be passed from ferret to human, human to ferret, but is not usually a serious condition, though, as is the case with humans, the young and the old are the most vulnerable, so it is always best to take precautions when your fitch goes down with a cold or 'flu.

If you have more than one fitch, it may be best to isolate the affected animal, though this is unlikely to help much, for it is very difficult to stop the spread of the common cold. It may be worth a try however, but you will also need to use a face mask and make sure that you wash your hands thoroughly after you have been in contact with the infected animal and before you handle any of the others. If you have young kits around or an old ferret, you should make every effort to keep the infection at bay, but if this is not the case it may be best to allow the cold to run its course and get it over with.

To be honest, out of many ferrets I have kept over the years, only one of them has gone down with a cold, and it was a very bad one at that, and I isolated him and he was over it and back to his old self after a few days of tender loving care. This was Jack, a hob ferret of the polecat type and he became very lethargic, sleeping most of the time he was ill, with matter in his eyes and a runny nose and a lack of appetite. I put him in a drawer which I left slightly open, but not enough so that he could get out, making sure of course that it was spacious enough for him to use the latrine, and I provided plenty of warm bedding, old towels and items of clothing which were suitable. He also sneezed a lot and I could often hear him sneezing away in his little isolation block.

Make sure there are always plenty of fresh fluids available to a sick ferret. Even though it may not eat for a couple of days, a sick fitch will survive no problem as long as it has fresh water. You may wish to add a little glucose to the water in order to help boost its energy levels which will help it to fight off the infection, but make sure that the glucose dissolves properly in the water before putting it into the bottle.

It is a good idea to invest in a small cage that can be used as an isolation unit whenever sickness strikes (which should be on a very infrequent basis I should say), or you could easily construct one of these yourselves, making certain, of course, that your fitch is not going to be lying on top of its droppings. You do not need to provide a great distance from the sleeping quarters to the latrine area, for your fitch may feel too ill to walk the distance required, but there should be a little distance, for the sake of hygiene and the clean nature of your fitch. The latrine area should then be cleaned out as often as possible, maybe three or four times a day whenever possible.

The symptoms of a common cold or 'flu can be similar to those of distemper. If your ferret has not been innoculated and it begins to display symptoms of what you suspect to be a cold, you must first of all isolate the animal, taking all hygiene precautions with washing and disinfecting, and take it to the vet as soon as possible, just in case. Of course, you may have a cold and could easily have passed it on. In that case, keep the ferret in isolation for at least two or three days and observe it closely. If it gets any worse, or has convulsions, or begins twitching, then get it to the vet for it is likely to have contracted distemper, rather than a common cold.

If you keep your fitch outside the home, then it is best to bring it indoors for the duration of the cold, as it will recover much better if it is kept in a warm environment. This is one good reason to invest in a small cage which can be used during periods of illness. If your fitch does not seem to be recovering after a few days, and especially if it fails to recover its appetite, then seek veterinary advice. If in doubt, go to a professional.

A sneezing ferret, of course, is not in itself indicative of an oncoming cold or 'flu, as ferrets, while in the course of satisfying their intense curiosity, a curiosity which undoubtedly aids the survival of wild polecats, will often get bits of dust or fluff up their noses and so sneezing will be quite a common occurrence when a fitch is exploring every nook and cranny to be found where they exercise. If a fitch is sneezing, but there are no other symptoms to be seen, then do not worry about this. Dogs too, will sneeze a lot when exploring with their noses. This is a most common occurrence, especially in areas where the vacuum cleaner cannot get – under beds for example.

Scouring (Diarrhoea)

A scouring ferret, one with diarrhoea, should be observed closely, for this may indicate something serious and other symptoms will undoubtedly follow if this is the case. Most forms of scouring, however, are not serious at all and cutting out all food for twenty-four hours should help to correct the problem, for you may have given your pet some food that was either a little too rich for it, or was turning bad. Make sure you provide plenty of fresh water in order to prevent dehydration, but it may be well worthwhile cleaning the water bottle thoroughly before you do.

When scouring occurs, analyze what you have been feeding and how often, for it may simply be that you have given too many eggs, or bread and milk, and, after resting the stomach for twenty-four hours, avoid giving such food for a good couple of weeks and afterwards provide these a little less frequently. If, however, the symptoms persist, then, again, it is the best policy, as with all other symptoms, to seek the expert advice of your vet.

Foot-Rot

The need for regular cleaning and disinfecting of a ferret cage, more especially those made of wood, is absolutely essential if you are to avoid your fitch contracting foot-rot, a most unpleasant condition, but one that is very easily kept at bay by good animal husbandry, something which I am sure you are determined to practise; the very fact that you have bought this book indicates that this is so. There is a danger, however that after you have been keeping ferrets for quite some time and the 'honeymoon period' has worn off, you may begin to slacken off a little in what otherwise has been diligent care of your pet. Damp, dirty cages can lead to foot-rot and the symptoms are swollen feet with scabs on them. Their claws may also drop out. An infected animal must be treated by a vet, and as soon as possible. Prevention is much better than cure, so keep up a good routine of cage and bedding box cleaning and you should have no problems, as I have had no problems, with foot-rot.

Tuberculosis

This is once again on the increase worldwide and so it is fitting to include this condition as ferrets, like humans, can suffer from this. The limbs can become paralysed and the body will waste away, with bouts of scouring (diarrhoea). If you suspect this illness in your pet, then isolate immediately and get it to your vet. This illness is highly infectious and should be treated as most serious. You should also visit your doctor and seek his advice if tuberculosis is confirmed in your pet.

Safety in the Home

Good health in a ferret will usually be maintained by good husbandry; a well balanced diet, fresh water at all times, clean cages, nesting boxes and litter trays, and clean environments for the use of exercise. We have already touched on certain aspects of keeping your pet safe while out at exercise, blocking holes which may be used as potential escape routes, putting wires out of the way so they cannot be chewed, preventing a ferret from getting inside electrical appliances such as washing machines etc, but I think it important to delve a little deeper into this subject as it is obviously closely tied in with ferret health.

To help prevent your pet getting stepped on and seriously injured or even killed, you may, as many owners choose to do, wish to fit a collar with a little bell attached which will hopefully make enough noise to make you aware of its presence nearby. To be honest, though this may work for a little while, you will soon get so used to the sound that you may stop noticing it so much, so always look around you and mark the spot where your fitch is. If you do fit a collar, then include a small identity disc too, as this may aid the return of your ferret should it escape and be lost.

A fitch will hate to have a collar on at first, but it will very quickly get used to it. You must make sure that it is tight, though not too tight for it to breathe properly, but without it being able to slip it over its head. Also, you must continually check that it is not becoming too tight as your ferret grows, adjusting, or replacing it,

A polecat ferret using its nose to explore. Ferrets will spend much time exploring their environment. This one is secured on a harness, so escape routes need not be blocked off.

when necessary. Of course, there is always a danger of a collar snagging on something, especially if it is too loose, but the danger is very minimal. If you have your ferret running loose around the home permanently, rather than just during exercise periods which you will closely supervise, then it may be the best policy to fit a collar, just in case of escape. Otherwise, as long as you take great care not to step on your ferret, then I believe it is unnecessary.

Because ferrets love to explore and sleep in dark places, you must be careful not to stand on something, a rug or an item of clothing for instance, which may prove to be a chosen hiding place for your fitch. It is the same with piles of washing, or laundry. Before putting these into the washing machine, make sure you account for the whereabouts of your pet, so that you avoid accidentally putting it into the machine and killing it. Also, whenever you go into your fridge or freezer, check that the ferret has not slipped inside unseen, for ferrets are surprisingly agile little creatures that can get up to all sorts of tricks without being noticed and this can sometimes put them in great danger.

Rocking chairs, or recliners, are also a hazard, for a ferret can be crushed if you are not careful. It is the best policy to keep reclining chairs in a permanent position when ferrets are around, and leave rocking chairs unused, or even in a room where your pet is not allowed access.

Plug sockets are another potential hazard, though it is most unlikely that a fitch will go poking around in one. Plastic covers can be obtained from any electrical or hardware store and these will provide complete protection when they are placed over unused sockets.

The safety of your fitch is also of paramount importance when introducing it to other animals. You may already have a cat or a dog and a young kit will be vulnerable at first, until your other pets get familiar with it and are used to having it around. You will need to take great care in this situation and closely watch your cat, thwarting any possible attempts at attack. If it is a fully grown dog that you are introducing your fitch to then it may be best investing in a muzzle. A dog can very quickly kill a ferret and terriers in particular are inclined to attack a creature of this sort, for terriers generally have quite a strong hunting instinct which can emerge in the form of cat chasing, or fighting other dogs with great enthusiasm, and, unfortunately, ferret killing. Use a muzzle for a while therefore, until both parties are familiar with one another and the dog has plainly accepted a new friend.

It is true that ferrets do have a strong hunting instinct too, but a kit which is brought up in the company of other animals, or even birds, will readily accept them as friendly company, though I would not recommend introducing adult ferrets to other animals without taking some precautions such as muzzling the ferret beforehand – if you are able to fit one of these contraptions that is. I have tried to fit a muzzle on a ferret and failed miserably, though you may have better skills than I have, so best of luck!

An adult ferret may attack a cat or a dog, and most certainly a bird, if it has not been brought up with other animals, so be careful when introducing these to your other pets.

It is far better, if you are intent on keeping a ferret and a dog or a cat, to purchase both when they are young and unspoilt and raise them together, for they will become playmates and firm friends in no time at all.

If you have a young baby, then under no circumstances should you leave a fitch running around a room where there is a baby around, *never*, even for a second, leave them alone together. Ferrets are predators by nature and the helplessness of a baby could trigger off these natural instincts. So always closely supervise a fitch around a baby. If a baby is toddling around the room, or is using a baby walker, then you may wish to put your ferret back in its cage to avoid any nasty accidents, letting it out again when the baby is asleep or has been put to bed. Again, with adult ferrets newly acquired and introduced to the home, use a muzzle for quite a few weeks when the baby is around, until you are certain that the ferret is trustworthy. These are sensible precautions, though I must stress that it is far better to purchase a kit in this situation. In this way you can get it used to being handled before introducing it to your baby, or young child. Ferret health and hygiene is a matter of good commonsense and animal husbandry.

Chapter Eleven

Breeding

Out in the wild polecats will become sexually active during the spring months and for very good reason. Food is more readily available at this time of the year and also the days are much longer after the dark, gloomy days of a long winter have passed, giving a polecat more time and more chance to catch food which will be desperately needed for the fast growing kits. Thus, when the days begin to draw out again and the weather softens somewhat, a female fitch will come into season, or heat, and she will then be ready to be mated. You will notice that the vulva will swell greatly during the breeding cycle, swelling many times greater than its normal size, and, once this swelling has finished, she will be ready to be served by a suitable hob ferret.

If you will not be breeding a litter, but you do not want to have a female spayed, maybe because you will be breeding from her the next year, then you need to be aware that an unmated female can, and probably will, develop complications. A jill ferret that is not mated will remain in heat for the whole of the summer and she could become ill because of anemia, or womb infections. She will become listless and very apathetic, usually losing interest in everything, including food and drink and thus she will often become dehydrated, losing weight and with a lifeless fur. This problem, which is usually fatal, even if you manage to get her to your vet fairly quickly, can easily be avoided by using one of two simple methods.

You can pay a visit to your vet who will administer hormone treatment that will bring the breeding cycle to an end for that year, or you can use a hob which has been vasectomised. This is not castration. A castrated hob will not be interested in mating a female. A vasectomised hob will, however, for he still has all of the necessary tackle, but will not be capable of providing sperm which is absolutely essential for pregnancy to occur.

You could use a vasectomised hob ferret belonging to someone

Kits at two weeks of age, bred from a polecat jill and a black-eyed white hob.

else, and you may choose to do so, but there are risks associated with such practices, such as the risk of spreading infection. So it is best, if you wish to breed some years, but not others, to keep your own vasectomised hob for the purpose of bringing your jills' seasons to an end. Be warned though, for a hob ferret which has had a vasectomy will still retain the strong odour which many find unpleasant and intolerable. So, if you do not wish to keep a hob which retains its natural odour, then it is probably best to have your jill treated with hormones prescribed by a vet.

A jill which has been mated by a vasectomised hob will usually have a false pregnancy and in many instances will come in heat again later in the year, thus you will need to employ the services of a vasectomised hob once again.

Your personal reasons for breeding ferrets may be varied indeed, but one thing is for certain – breeding for profit is very unrealistic and the modest rewards are just not worth the effort involved. Even if a pet store will take all that you can supply, it is unlikely that you will get a very good price for your young stock and you will certainly not make a living breeding ferrets. Not only is this form of breeding unrewarding financially, but you will also have the worry of wondering if your youngsters have gone to good homes. If, however, you breed just for yourself and maybe a few friends, then the breeding of ferrets can be most rewarding indeed, for, in years to come, you will have enjoyed the company of several generations of ferrets whose parents and grandparents were owned by you also.

You may also become serious about showing ferrets, or even racing them, and it is most satisfying to produce your own winning stock from your own winning stock. Indeed, the offspring from a top winning ferret, male or female, will undoubtedly be very popular, though, of course, there are absolutely no guarantees whatsoever that a top show winner will pass its desirable qualities onto its youngsters. The only danger with breeding exclusively to produce fine show animals is that, in attempting to produce that 'perfect' specimen, you are very likely to need to breed in quite large numbers and the ones which do not measure up will have to be found good homes. So, though breeding for show purposes is a good idea and a very rewarding practice, breeding obsessively to produce that incredible animal which wins top honours nearly every time it is exhibited is not a good idea and can only do more harm than good.

For instance, in order to determine which in a litter of kits is going to turn out the best of the bunch, you will need to keep the whole litter until they reach adulthood and thus their full potential. If you breed from more than one jill, then you will need to raise more than one litter. This can become expensive and very time consuming and, what's more, you cannot then give an individual ferret enough attention and training to become a well socialized, housetrained fitch which will be a delight to own. You will then have to find homes for the rest of the unwanted litter. Of course, you may develop a knack, an instinct, call it what you will, of picking the best of the litter while they are kits, but you will still have the task of finding homes for the others and this can be difficult when you breed in large numbers. So my advice is to breed for yourself and any friends or family members who wish to take up ferret keeping.

The Mating

When the jill ferret has come into season proper, the vulva being fully swollen and you may notice a little discharge too, it is time to introduce her to the hob who will also by now be ready to mate, for even the hob comes into breeding condition during the early spring. If you do not keep your own hob ferret, then you will need to use one belonging to someone else, possibly a fellow ferret club member, but make certain that both your jill and the hob belonging to a friend have been vaccinated particularly against distemper, and make certain also that the owner of the hob is diligent about good animal husbandry, for you do not want your jill catching mites, or even mange, from a neglected hob.

Take the jill to the hob and make sure they have plenty of privacy and it is best to confine them in a cage which is quite roomy, for the hob will grab your jill by the back of her neck and will drag her around and generally treat her in a very rough manner indeed. Do not be alarmed by this and do not attempt to interfere in any way whatsoever, for this rough handling will ensure the maximum chances of conception, so leave them to it and go off and do something else, for your jill will protest and may even scream with pain, or rage, while being mated, and this may cause you distress. Do not worry, for this is totally natural and your jill will come to no harm, except for a few bite marks on her neck.

Leave the pair together for a good few hours, maybe even for a full day, and then separate them. You may wish to repeat this procedure a couple of days later, but, to be honest, this should not be necessary, as conception should now occur without any problems.

After the jill has been with the hob for a few hours, it is advisable to check the back of the neck for bites. Using boiled water which has been allowed to cool and is slightly salted, clean these bites with cotton wool dipped in the water two or three times a day until they have dried and scabbed over. If these bites become very red and swollen, then they have become infected and a visit to your vet will be necessary, for your fitch will probably need antibiotics, though this is most unlikely I have to say as bites received while mating, as long as you do not leave your jill with a hob for days on end, will rarely give you any problems, particularly if you give them a good clean with the salted water as soon as she and her mate are parted.

Pregnancy

Over the next few weeks the vulva will return to its normal size and the actual length of the pregnancy will be six weeks (forty-two days), give or take a day or two. There is no need to increase greatly the amount of food you provide, but make sure there is always some food in the dish throughout the day, thus allowing the mother-to-be to take food at her leisure. Springtime in the wild is the time for nesting and raising young birds so eggs will now be on the menu of wild polecats. Make sure you provide raw egg in the diet of a pregnant jill and one which is raising young, but do not give too many, lest her stomach gets upset and she begins scouring.

It is best to put a pregnant jill in a cage on her own. True, many ferrets will do fine when left with others, but there is always the risk of something going wrong and the kits could be eaten, or the mother may attack other ferrets if she is left with them. However unlikely this is, it just isn't worth the risk, so select a cage especially for the purpose of raising the litter and leave her on her own, settling her in at least two weeks before she is due to whelp (give birth).

After the first four weeks of pregnancy you will begin to notice the mammary glands swelling and the nipples becoming more prominent, the abdomen may swell enough for you to tell that the birth is now very near. Your jill will also spend quite a bit of time making a suitable nest for her youngsters, so make sure you provide plenty of material for her to use.

For those of you who allow your fitch to have free run of the home on a permanent, or even a semi-permanent basis, then it is still the best policy to confine them in a cage a couple of weeks before the birth is due and for the duration of raising a litter. You could use a sturdy cardboard box and many may do this, but I would not recommend this for raising a litter. If an accident did occur, a litter of kits could easily be crushed in a cardboard box and it will become soiled much more quickly than a cage which will also offer more protection to a growing litter and their mother. If you do allow your ferret free run of the home, she can then have her young literally anywhere she chooses, or she may even have them where she can hide them away from you. Also, when the litter begin moving around, they will not be housetrained and, what's more, you will find it impossible to even attempt housetraining a full litter, so confine your jill until the litter is old enough to be without their mother. This will do her no harm whatsoever and, once the kits are a bit older and can be left for a period of time, you can bring out your jill for exercise, as long as you do not keep her from them for very long.

By the sixth week of pregnancy, you may be able to feel the kits moving around inside her, but be gentle when feeling for babies. If she is having a phantom, or pseudo pregnancy, your jill will still go through all of the motions of being pregnant and, after the due date has been and gone, she will act as though she has a litter of kits – lying on her side as though she is suckling a litter for example. After the pregnancy period, she may come in season again and you can have her mated once more, if you so desire. Remember to wait until the vulva is fully swollen before having her mated again. After the first swelling begins, a jill will usually be ready to be mated between two and three weeks later. A jill could suffer injury if she is mated before her vulva is fully swollen. At best, she will suffer much pain during mating and all for nothing, for a ferret will not conceive when she is mated too early.

Newborn Kits

If your jill is having a genuine pregnancy you may be able to feel the litter inside during the last stages. Your jill will probably sleep a lot more as her time approaches, or she will just enjoy being in the nest awaiting the imminent birth of the litter which usually occurs,

A nest full of kits at two weeks of age. Shredded paper is ideal for use in the nesting box and these kits look warm and comfortable. Note the dark polecat colouring beginning to show on some of the kits. The others will be white.

for some extraordinary reason, during the night. You will soon know if your fitch has given birth in the night, for she will not be out of her nest to greet you in the morning. Make sure that the nest box has a lid which you can lift off and just have a quick peep inside. There is no need to sex, or even count them, at this time, for it is best to leave her to it for the first few days, until she is more familiar with her new litter and has accepted your visits to her cage as being of no threat to her young. Wild polecats, and other members of the ferret family, will vigorously defend their young from any threat and so you must respect the privacy of the mother and young during the early period especially. If a jill is badly upset at this time, she may even turn on her litter and kill and even eat them. With this in mind, you can see why it is important to keep her and her new family quiet for the first few days. You will still need to provide food, fresh water and a clean litter pan, of course, but, after the initial first peep inside the nest, leave her to it for maybe a week, before having another look.

When you peep inside you will see tiny little pink things that are both blind and deaf and they will be totally dependent on their mother for the first few weeks of their lives, continuing to suckle even after weaning, though, as they get older, she will become less and less tolerant of them suckling as their growing teeth and claws can cause some discomfort, in much the same manner as bitches will become less tolerant of their puppies.

Care of the Whelping Jill

As the kits grow so they will take more and more out of your jill, so make sure she has plenty of food and fluids. Some think that providing milk will help a jill to produce more milk, but this is not so. If you do wish to give her milk, then water it down to reduce the risk of her scouring. Also, make sure that the diet is well balanced, high in protein and with enough fats and carbohydrates. She will lose quite a bit of weight at this time, so make sure she has a constant supply of food in her bowl and, of course, plenty of fresh, clean water. Your jill may look a little ragged at this time too, for she will undoubtedly shed her winter coat in readiness of growing her summer coat, the period betwixt leaving her looking well below her best. Do not worry, once that coat has grown again and the litter are fully weaned, she will soon be back to her old self.

Feeding the Kits

The kits will soon become covered with white fur which will darken to a muddy sort of grey by the age of three weeks when the eyes will begin to open. They will also begin eating solids at about this time, having already begun leaving the nest, much to the annoyance of the jill who will try to keep them contained within the safety and security of the nesting box. Kitten or puppy food will do fine for feeding to kits, though you must supplement this with some raw liver cut up into small pieces.

Wild polecat jills take their kill to the nest and she will allow the youngsters to tuck in and help themselves, but she will help them out by opening up the carcass so that they can get to the meat much more easily. With this in mind, it is unnecessary to mash food for them, though I would cut raw meat into smaller pieces for them during the very early days of weaning. Raw eggs are also good for kits, as long as you do not feed too many.

Care of the Kits

Albino ferrets, of course, will remain white, but the polecat type will get darker and their markings much more prominent as they grow. Once the eyes have opened at around three weeks of age (some may open earlier, while others may open later) and if it is a white fitch, the pink eyes will soon be visible, but you may be fortunate enough to breed a white with black eyes, a most attractive animal. If you do not wish to keep this one for yourself, then I am sure you will not have a problem finding it a good home as this type is very appealing and pleasing to the eye.

You should give the nest box a good clean out about once a week before the birth of the litter is due and then leave the jill to it, for she will make a nest in readiness for the impending birth and she should not be disturbed once she has done this. The actual birth can be a little messy, but, be assured, your jill will clean up after each kit is born. These can be born moments apart, or hours apart, it all depends on the jill but it is usually all over in quite a short time. Once the birth is over and apart from an initial peep at the slender, wriggling, naked (though they look pink and naked, they are actually covered in a very thin white down) little creatures, leave her alone except for the cleaning of the litter tray etc. You may begin cleaning the whole of the cage, minus the nest box of course, after five or six days when she has settled and is now sure you are no threat to her kits. You can also have a longer peep at the litter, but I would not handle them for a few more days, just to make sure, and then be careful that your jill does not bite you. This is very unlikely but it is better to be cautious when first handling them. If, when you are lifting them out, the mother gently takes them off you, then allow her to do this and leave it another day or two before you try again, for you will upset her if you continue to attempt to take them from her when she has politely let you know that she is not happy with the situation. Some mothers will let you handle their kits with no objections at all, while others will be most reluctant to allow their kits to leave the nest. This is a natural instinct for a wild polecat jill cannot protect her young if they are wandering all over the place. She will keep them close to her to prevent them becoming fodder for predators, and this is why your ferret will keep taking her young back to the nest even though they are keen to do a little exploring of their surroundings.

The mother will keep the nest clean for the first three weeks but when the kits begin eating, they will make a real mess and so you now need to clean the nest out at least once a week, but checking at the end of each day for leftovers and removing any that you find. You do not want bad food left in the cage as this could make your kits seriously ill, or even kill them, especially when it is warm, so be diligent in this regard. It is good to do this while the litter is in the nest, for this will make them very familiar with hands and fingers and human scent and this can only help to socialize a youngster, making it into an ideal pet. In fact, from the age of three weeks onward, once the eyes have opened, it is a good idea to handle and play with them on a regular basis. In this way, by the time they are ready to go to good homes, they will be well used to being handled and will be far less inclined to bite, though they will still 'mouth' as they are still gaining their education about everything around them, whether it be edible or not!

If the nest begins to smell, then you may wish to clean it out long before the kits have reached three weeks of age, but do not even attempt to do this until the kits are at least a week old, lest the jill is upset.

The size of the litter is usually around six or eight kits but there could be as few as one, or as many as twelve. Do not worry, your jill will have no problems with raising a larger litter as long as you help her out by providing enough food from the age of three weeks onwards.

Though in most cases the whole of the litter will do fine, there may be one of the litter which is struggling a little and is growing much slower than the rest. This is known in livestock circles as the runt of the litter and it may need a bit of extra care from you. It is a good idea to feed this one separately so that you can gauge just how much food it is getting, for the runt will usually eat much more slowly than the rest, and so it will get little food. Feeding it separately several times a day will ensure that it is getting enough food to grow properly, though it may always be a little smaller than the rest. The runt may also be a little on the shy side, so the extra attention it gets will help it to become a more sociable creature.

When the litter has reached about four weeks of age, it is a good idea to allow the mother time out from her family and you can do this by allowing her to exercise away from them, though you must not keep her away for too long as this may distress her. Their teeth and claws will now be causing her a little discomfort and quite a bit

Polecat kits belonging to Derek Webster, the darkest I have ever seen. There are variations on all fitch colourings.

of annoyance, so she will need this time away and on a more frequent basis as the litter grows. By the time they reach six weeks, they should be fully weaned and the mother can be taken away from them on a more permanent basis, though you can still allow them periods of play together. The kits will probably attempt to suckle, but she is unlikely to allow them much by this time and will often get out of their way. At eight weeks they can leave the nest altogether and are ready to go to their new homes – homes which you will no doubt have made sure are good homes, especially if you breed for yourself and friends or family.

Problems in Breeding

Though it is quite rare, a fitch giving birth may encounter problems. A litter will usually be born unattended and I would not recommend interfering in any way, but, if you are around when she goes into labour, and it becomes obvious that she is in trouble – she is pushing but nothing comes out and she is obviously in distress, then call the vet in. If a kit is hanging out, but is stuck, then very gently, if your jill will let you, help to ease the kit out, but *very gently*. If a kit is born and the mother does not attend to it, then get a rough cloth and *gently* break the sack from around its mouth and nose, making sure the cloth is damp, for this emulates the tongue of the mother and the kit should soon begin breathing and wriggling and the jill will now usually take over. Some though, will eat all of their babies and this can be most distressing for the owner. A wild polecat will kill and eat her kits if (1) there is something wrong with the litter (2) if they are severely threatened by a predator (3) there is something wrong with the jill and she knows she just isn't capable of looking after them and (4) there is a serious shortage of food and the litter would starve to death. As long as you provide plenty of food and respect her privacy during the early stages of motherhood especially, then there should be no problems in these two areas. If your fitch does eat her kits then it may be because there was something wrong with either the litter, or herself. If she seems fine to you, then it was probably the kits, but not necessarily, for some jills, as is the case with dogs and cats, are just bad mothers and will either eat, or totally neglect, their youngsters.

Your jill may come back into season and you may wish to have her mated again, or you may have to wait until the following spring, but, if she eats her kits again, or even abandons them, then I would not breed from her again. I would have her spayed as soon as possible and breed from another, more suitable jill.

If a litter of kits is abandoned by their mother, you must quickly find them a foster parent, or they will soon die. Another jill rearing youngsters will do the job fine. It is a good idea to have two jills pregnant, having them mated around the same time, so that one can be used as a foster parent should anything go wrong. If you cannot foster a litter, then it is best to have them humanely destroyed. Of course, you may not wish to breed from two jills just in case one of them abandons a litter, or becomes too sick to look after them, so you may wish to ask a friend who is intent on breeding, or a fellow ferret club member maybe, if they would have their jill mated at around the same time as yours. Alternatively, if you do experience problems with your jill and you know of someone else who has a litter, then you could ask them if they would allow their jill to become a foster mother to your kits, though, if a jill fitch already has a large litter,then this may not be possible, for you do not want to put in too many foster kits, lest she cannot feed them all. A litter of twelve is large enough, though some jills will raise as many as fourteen successfully.

To be successful at fostering a litter of ferret kits, you need to be a little on the cunning side, for, if the mother has any idea of what is going on, she may well kill the foster kits. Distract the jill by feeding her well away from the nest box. If the litter is newborn, she will most likely carry her food back to the nest, so give her some bread and milk, or some other food she cannot carry, so that she remains at her food dish for a while. Whilst she is there, put the foster kits into the nest and gently rub them with the bedding and rub them against the blood kits, so that you cover them with scent which is familiar to the jill. In this way you will hopefully succeed in tricking her and she should have no idea of what has occurred, for ferrets cannot count and she won't realize that her litter has increased. If she already has a large litter and you can only place four out of eight kits with her, then, if there is another jill available, the other four can be fostered with her. If there is not another jill available, then the only option is to put the rest down.

Mastitis can cause problems for a jill nursing young kits and her teats will become hard and swollen and very sore, so you must seek veterinary aid immediately, or the kits may be unable to feed and they will die. The jill will also deteriorate if this condition is left untreated.

Milk fever is another condition which can badly affect a nursing mother and it usually strikes after a few weeks of caring for the kits. This is caused by a lack of calcium and if you provide a well balanced diet, then this should not be a problem. Eggshells crushed up into a powder and sprinkled on food will help to provide extra calcium during nursing, as will a food source, such as a complete ferret food, or a carcass of a rabbit, that is rich in calcium. Fits and paralysis are common symptoms of this complaint and immediate veterinary aid is essential, for your jill will need a calcium injection and a proper diet containing plenty of calcium if she is to recover, for this condition can be fatal.

When breeding animals of any kind it is essential to keep a close eye on both the mother and her babies, making sure that they remain in good health. Although we have seen that there can be problems when undertaking a breeding programme, these problems are rare indeed as long as you are diligent about providing a good balanced diet and a clean environment, along with respect for the privacy and protection of the mother during the earlier stages of raising her young especially. So please, do not be put off breeding your ferrets, for this can be a most rewarding experience and it is very satisfying watching your youngsters growing and seeing their personalities developing. You may even notice qualities, good, bad, or even amusing, that have been inherited from either the mother or the father.

If, after successfully raising a litter and placing them in good homes, your jill comes into season once again that same year, then I would not recommend breeding from her again. This should not occur, but it may do in some cases. If your fitch is intent on breeding again that year, then either have her treated with hormones, or have her served by a vasectomized hob. If she has eaten her kits and you are intent on a litter from her, then in this situation it is all right to have her mated again in the same year, but do not make a practice of it, for having too many litters can do more harm than good to a fitch. It is also good practice not to breed from her every year, though this is a choice you will have to make.

Chapter Twelve

Exhibiting

Though ferrets are still very popular as working animals in countries such as Britain, Ireland and Australia (rats and rabbits are controlled out of necessity in such countries and the ferret is very useful in this regard), in many other countries as well as these three, increasing numbers of people are turning to ferret keeping and are enjoying their company as pets, as opposed to working animals, or as show animals, for ferret shows continue to grow rapidly in popularity and this can only be good for the reputation of fitches which often get unjust publicity.

Exhibiting ferrets can be very rewarding and satisfying and is a grand day out, as long as you do not take the showing side of ferret

Exhibitors with a selection of albinos.

keeping too seriously, for that could lead to disappointment and you will become disillusioned with the show scene in no time at all. If you view a day out at a show as an opportunity to meet like-minded people and a place where you can enjoy a wide variety of different types of ferret, maybe even as a place where you can improve your knowledge of, and thus the care of, ferrets, then you will have a great time and you will undoubtedly enjoy showing, as long as you view it as a bit of fun. Winning is just a bonus that will only enhance an already enjoyable experience.

If you set out with big ideas about winning every time you exhibit at ferret shows, then you are going to end up very miserable indeed, for a judge may not have your idea of what makes a 'perfect' fitch and, indeed, what one judge views as a champion, may not even get placed by another judge at another venue, so set out to have fun and then, if your ferret gets placed, or even wins the show, then it will be a far more pleasant day out. Not that you have to keep your expectations low, for each individual owner will no doubt view their fitch as a potential winner, so there is nothing wrong with hoping for a place, or even the championship, but expecting to win and focusing on winning alone will only lead to frustration and extreme disappointment, totally ruining the day for you.

The first thing to do is to find out when and where ferret shows are held. Though there may be one or two shows in your area, you must be prepared to travel if you wish to exhibit on a regular basis. Again, if you travel around one hundred miles to a venue and you are absolutely intent on winning the show and nothing else will do (there are a surprising amount of people who do have this attitude towards showing, whether it be ferrets, dogs, or some other animal), then, if you do not win, can you imagine what a burden that journey has now become? A long trip home is now ahead, during which time you will have plenty of opportunity to dwell on your bitter disappointment. I have seen it happen time and time again, the angry exhibitor storming out of the ring, with some even hurling abuse at the judge as they go, simply because they set out to win and nothing else will do for them.

Ferret clubs and welfare societies will often stage shows in order to provide a fun, sociable outing for their members, to promote the true image of ferrets and especially their value as pets, and, of course, to raise funds. If you are a member of one of these, and I recommend that you do join your local club, especially if you wish to exhibit, then you can find out from them when and where a show will be staged. It is also a good idea to look in any publications which advertise such events. In Britain and Ireland we have *The Countryman's Weekly* which advertises these shows.

Show Preparation

Apart from finding a venue and planning your route to the showground, there is little else you need to do in order to prepare your fitch in readiness for exhibiting. You may wish to bathe it the day before but, as long as the ferret has a glossy, healthy coat, there is little benefit in doing so, though a good grooming, just to get rid of those dead hairs, is a good idea. This can be done with a soft brush on the day of the show, but is not really necessary at any other time. Dead hairs will be dry and a brief grooming will take these out of the fur and may improve your chances, if only slightly.

Never stick cotton buds inside the ear to clean them. If they are really mucky inside, then get your vet to deal with this, but if there is only a little bit of wax, this is normal, so leave well alone, for no judge who has any inkling of what he is looking for will knock a fitch which has a bit of wax in its ears. If he/she does, then they should not be judging. You can safely clean the outer flaps of the ears though, if you think it is necessary.

Polecats living wild cover tremendous distances searching the countryside for food and during the springtime breeding season cover great distances looking for a suitable mate and this keeps their claws quite short because of the wear and tear of crossing what is very often rough ground. A ferret's claws should not therefore be long and sharp. Clippers used for dog and cat claws are suitable for this purpose, or even nail clippers will do, but only take off the sharp ends. You must avoid cutting the vein inside the claw, as this will cause pain and a little bleeding and may make your fitch a little lame, something that will go against it in the showring. Clippers which are suitable for use on cats' claws, or even normal everyday nail clippers, will do the job well enough and, as long as you do not cut the quick, this will cause no pain or discomfort at all. Because ferrets are natural predators, they must have some length on their claws as this will assist a wild polecat whilst hunting for food. For instance, if a rabbit jams itself into a stop-end inside its burrow, with its back-end towards the predator, the polecat will scratch at the rabbit's backside in order to make it move so that the fitch can get to its head and thus be able to finish its quarry quickly and cleanly. So a judge will take such matters into consideration, for he must look for points that will aid a fitch if it was used for hunting purposes, or if it was in its wild state.

The coat of the fitch to be exhibited should be clean and glossy and the growth should be even throughout, so you must wait until both the summer and the winter coat has grown properly, or your ferret will do poorly otherwise, for no judge would award prizes to a fitch with a poor coat. Bald patches will also go against it and if your pet has these even after moulting has occurred it is best to get your vet to check it over, for mange can produce bald patches, as

Andrew Hogg with his winning ferret 'Rambo'.

131

can other conditions such as a poor diet. The colour and markings are really only a disadvantage, or, indeed, an advantage, according to a judge's preference and tastes, though a hob ferret of the albino variety which has its white fur tainted to a dirty yellow (something which can occur with sexual maturity), may do very badly when up against an almost pure white castrated male. So the coat must be healthy and the colour and markings are totally in the hands of the particular tastes of the judges.

The overall shape of a ferret should be well balanced and proportionate, that is, it should be pleasing to the eye, the ears, legs and body length of the correct size, the head being just right, not too small, or too large, in comparison to its body. The body should be quite lean, but without the animal being underweight. Surplus fat will not do much for a fitch in the show ring, so you must find the correct diet that suits each individual and make sure that you avoid giving too many treats, especially such things as chocolate, for these will quickly pile the pounds on if you are not careful. A fitch which gets plenty of exercise will always look healthier than the one that is stuck in a small cage for most of the time, so be certain to schedule enough exercise time, for this will keep the animal looking lean, fit and with good muscle tone. This will also help to keep the coat looking right, for a fit, healthy ferret will usually have a glossy, healthy coat. If you exercise your pet outside at all, maybe in a yard where there are concrete or stone flags, this will help to keep the claws worn down so that clipping becomes unnecessary. Polecats hunting over rocky ground keep their claws worn down naturally, for they will receive much wear and tear as they are regularly used for getting a grip on often slippery rocks.

The fur, apart from being glossy, should also be dense and soft to the touch. If it is hard, then it is dry and is probably lacking in natural oils. Providing a feed with natural oils, or including cod liver oil in the diet, should remedy this problem in no time at all. The fur on the tail should be quite bushy. The coat should have plenty of life in it, it should not be dull and lifeless, for this will go against a fitch in the ring and you will no doubt be advised to get the animal checked out, for an unhealthy coat usually indicates an unhealthy fitch.

The eyes should be clear and bright and the teeth should be clean, the jaws neither undershot nor overshot, a fault which can be hereditary in some strains, especially where inbreeding is practised, that is, the mating of related ferrets in order to keep desirable traits, either for hunting, or for show, in the offspring. Unfortunately, the undesirable traits can often show above all others, so inbreeding is not always a good idea.

The exhibit should be alert and very active, full of life and vitality and radiating health and fitness, qualities which are essential to the hunter of rats and rabbits, and especially of the wild polecat, for an unhealthy, unfit wild polecat would surely starve to death in no time at all. A fitch should also be very tame and easily handled, even by a stranger, so it is essential that you get your ferret well handled by both yourself and any willing visitors to your home. The more socialized a show ferret is, the better, for you will have no fear of the judge being bitten by your 'little darling' while he or she is handling it, for that would certainly go against you and a biter is not fit for the show ring at all and is best left alone. If a judge wishes to handle your fitch, then he should be able to trust your word that it will not bite. Ferrets used for rabbit and rat hunting, if handled properly, will be just as tame and easily handled as those kept for their very desirable qualities as pets.

If you look after your fitch well, being diligent about providing a safe, clean environment for it to live, a good, well balanced diet and plenty of exercise and play, and then paying some attention to a few smaller details just before the show, such as those already mentioned, then you are doing your utmost to present a fine animal on the day. From then on, to be quite honest, it is a bit of a lottery and the judge, at the end of the day, will probably pick the one that he would most like to take home, for a ferret can be a champion at one show and then, the week after, it may not even be placed. If a judge is more inclined towards albinos, then there is a good chance that an albino will take the championship. If he prefers the polecat variety, as I do, then there is every chance that the winner will be of this type, for tastes vary greatly and what one judge considers a top winning champion, another judge may not even give a second glance. This is another very good reason for not taking showing too seriously. This is a good thing, I suppose, for if all judges had the same tastes, then the same ferrets would win time after time and no one else would get a look in. Of course, sometimes an outstanding example is produced in a litter and these can become champions several times over, but, generally speaking, the decisions of different judges will differ greatly, so you are always in with a chance.

Judging

Those who are entrusted with the responsibility of judging should be from among those who have many years experience of keeping ferrets, and many judges will also have experience of exhibiting, though it is not necessary to have as a judge one who has taken several championships. It is far better to have one who is familiar with both winning and losing. Judging is a thankless task and you will almost always upset someone, so a judge, apart from knowing his subject inside out, must also have a very thick skin, for criticism is sure to follow every decision he makes. On the other hand, the exhibitors must recognize that the decision of the judge is final and that it is pointless (not to mention a major irritation) to go running

Making sure this polecat is of the right sort.

Are those eyes and
ears clean?

to show officials in order to complain. It is far better to accept these decisions with good grace and try again at the next show. Having said this, sometimes judges do make some baffling decisions and you can often hear the crowds mumbling their disapproval around the ringside.

Judges should not be appointed to officiate at local venues for fear of complaints about the picking of friends' ferrets, rather than the best of the bunch. This can and does occur and I have seen this happen on more than one occasion, particularly at some of the dog shows I have attended, so it is best that a judge travels out of his own district in order to avoid this. One has to sympathize with a judge in this situation though, for it is difficult not placing the animal of a friend and this can make the situation a little awkward, so a judge could ask his close friends not to exhibit at his show, or he could

135

warn them beforehand that he will be judging the animals, rather than their owners, so they should not be surprised if they do not get placed. I have judged with friends in the ring and it does make one feel very uncomfortable when one does not place the animal belonging to a friend. Judging is, however, very rewarding, so do not miss an opportunity to judge once you have become more experienced and your knowledge of ferrets is such that you know what you are looking for in the ring.

Show Essentials

You will need a spacious carrier for when you travel to shows and it is a good idea to take some cleaning equipment and fresh bedding, just in case your fitch uses a corner of the carrying box as a latrine. This will probably happen when the ferret is in the box for a while during a long journey and whilst waiting for the show to begin, though, if possible, the organizers should provide a suitable area where ferrets can be allowed some space and can be fed and given a drink. Always make sure that you have plenty of fresh water avail-

This fitch waits patiently for the judge to finish.

Travel to the show – a wooden box which is ideal for travelling by car.

able and a supply of fresh food too. Give just enough to take the edge off your ferret's appetite, but not so much that it looks bloated and dozy, for this may take the edge off its alert, active personality. This would undoubtedly go against you.

It is also a good idea to take along a plentiful supply of tissues and some clean water in a small spray bottle, together with a suitable cat brush or comb, just in case your ferret gets its coat soiled whilst travelling inside the carrier. This is well worth doing, for it only takes a few minutes to get your fitch clean again.

Be careful to keep the carrier always within view and, while you are in the ring exhibiting, be sure to have someone reliable watching over your fitch, lest a thief pick up the carrier and makes off with it. Most ferret owners are honest, caring folk of course, but there is always the risk of thieves at any public gathering, so always be aware of where your fitch is. Also, make sure the carrier is not placed in direct sunlight, for this could prove fatal. Cages too should not be

The winner; Kirsty Brunt with Solo.

situated in direct sunlight. A distressed heavily panting fitch is one suffering from heatstroke. Immerse it in cool water (not cold or freezing which may cause shock) for a few minutes at a time until it recovers. Likewise, at winter shows, make sure that your pet is both warm and comfortable, for the cold, just like the heat, can kill if you are not careful in this regard.

I know I have mentioned this in a previous chapter, but it is essential that you do not leave your pet inside the car on a warm day, especially when it is parked in direct sunlight, for the car soon becomes very hot and this can very quickly kill any animal left inside. And remember, the sun moves across the sky during the course of the day and your car may end up being parked in direct sunlight at some point during the day, even though it started off in the shade! So think carefully before leaving any animal inside your car during the warmer months, for many countries will award severe penalties to those who do so. If you do not take this point seriously, then try sitting inside your car on a warm day and see how long you can stand the heat.

Remember to enjoy your showing, no matter what the outcome!

Chapter Thirteen

Ferret Racing

Another good reason for attending ferret shows is that they very often stage ferret racing as an extra attraction which usually proves very popular indeed, for much fun can be had while both watching, and participating in this event, an event which seems to be rising in popularity each year. It is a competition that is much more easily decided than the actual showing of ferrets, for the winner is clearly seen by all, along with the other placings too. Indeed, some, possibly because of this very reason, shun the show world and concentrate on racing instead.

Ferret racing began in the United States of America and soon became increasingly popular in a good number of countries including Britain, where the sport has now become very well established.

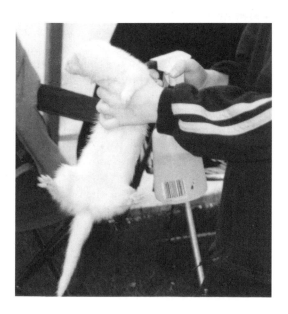

Preparing for the race.

Increasing numbers of shows put ferret racing on their list of events and thus increase greatly the funds they raise for ferret welfare and ferret clubs, all of which are trying to promote the excellent qualities of ferrets and their widespread suitability as pets, with much of their work promoting the proper care of this charming little fellow.

Legend has it that oil workers used ferrets to pass wires through pipes and that, during a lull in the day's work schedule, they would take bets on which ferret would be first through the pipes. This story can only be speculation for when I first began keeping fitches when I was a lad, many stories were told of electricians using ferrets to pass wires through pipes and under floorboards etc, so it is difficult to fathom the truth where legend is concerned. One thing is for sure however and that's that ferrets are curious, very active creatures, naturally drawn to dark passages, and have an inherent trait passed on from their wild ancestors which means they need to explore dark tunnels to find both food and shelter, and it was inevitable that one day men would use them for a sport of this kind.

Lengths of pipe with gaps at intervals so that the progress of the competing animals can be followed by the onlookers, are what is used for the racing of fitches, and there is much you can do to train your pet so that it is able to compete with the other animals at a decent level. I am sure you will really enjoy ferret racing and it is well worth the effort it takes in order to prepare. For instance, where the onlookers' gaps are in the pipe may stop a ferret in its tracks, for curiosity can take over and a fitch may spend quite a bit of time

Clear piping is better than boxes for breaks, as fitches are very curious and will explore this box-type break, as this albino is intent on doing.

Getting ready to enter the ferret.

An albino ferret emerges and heads for the finishing line.

exploring each gap in turn. You can easily combat this problem when exercising your pet at home, or in the garden.

Lengths of pipe are cheap enough and you can easily cut these into smaller sections so that you can include gaps in your tunnel system, thus imitating those used at racing events. Put a treat at the far end of the tunnel system, maybe a small piece of liver, or something similar (you can only usually give one or two of the manufactured treats per day, so liver is a far better option) and enter your fitch at the opposite end to where you have placed the treat. Make sure this is done before the main mealtime, for your fitch will be far more eager if this is done while it is hungry. It would undoubtedly still travel the length of the tunnel even with a full stomach, but it would be far more sluggish than when it is entered whilst still hungry.

Once the ferret has caught on to the whole idea and it realizes that a treat awaits its arrival at the other end, then it won't spend a lot of time exploring each gap it comes to, but will instead get on with the job in hand and will be out the other end in no time at all. After a few goes at this, your fitch will no doubt be a made racer but, in order to keep it keen, you must do this quite regularly at exercise times maybe no less than once a week.

If you are a member of a ferret club, or you have one or two friends who also keep them, then it is a good idea to get together regularly and set up your own little racing circuit. This can be great fun and is a good way of cementing friendships, as long as the results are taken very lightheartedly, for problems can arise if one or more of the owners is too competitive. Remember, these occasions are really training sessions for your respective ferrets in order to prepare them for the real thing, though you will more than likely find that many of those who are there for the racing, will take it very seriously indeed. In fact, some try to breed slimmer, more agile fitches just for the purpose of being more competitive at these events. This is ludicrous. We have seen the damage that has been done to dogs for the purpose of finding a top show winner, so the breeding of any animal just for show or racing points, is unnatural and is a bad practice. They should be left to produce offspring of the correct size and proportions as they are fitted for the natural world, not to suit some man-made image of so-called 'perfection'.

Racing is just another joy to be had by the keeper of ferrets and is, just like showing, a great way of meeting like-minded people and their great variety of fitches.

Chapter Fourteen

Organizing Ferret Shows
and Races

Showing your ferrets and racing them can be great fun and a source of much satisfaction and actually organizing these events, though hard work, will bring more satisfaction, though possibly a little less fun. It is possible to organize such an event on your own, but this is unadvisable, for it is quite hard work and there is much to do, so it is far better to form a small committee made up of family and friends and then the different tasks can be delegated between you all, making the load much lighter and hopefully ensuring better efficiency. If you can only round up one other person to help run the event, then this is much better than doing it all on your own, though, of course, on the actual day of the show, you are going to need a few helpers, to act as stewards etc.

An assortment of ferrets at a show. They are very inquisitive and are nearly always on the move.

The Venue

The first thing to do is to find a suitable venue for holding the event and there are a few things to be considered before you make a decision as to where the show will be held. For instance, the show-ground must be easily reached and easily found by those who are from out of town and are not familiar with the area. To help with this, you may need to make and put up one or two temporary road signs which will need to go up on the morning of the show day, and then they will need to be taken down again once the event has finished, so you will need someone to do this. Also, the next consideration is parking space. There must be plenty of room for parked cars in and around the actual showground, for some may bring many ferrets to the show and they will not wish to lug carrying boxes for any distance. Some shows do have plenty of ringside parking and this is very desirable if you can get it, though venues of this kind are very few and far between. The next thing to consider is cost. You will need a venue that will benefit them financially, that is, the

Rosettes and trophies for showing and racing.

144

owners, allowing you to stage your show on their premises. A country inn, for instance, would benefit greatly from the extra drinks and food sale which would ensure after a crowd of people descends on the place. Or, if there is a country show in your area, maybe an agricultural show or possibly a dog show, you could approach the organizers of this event and ask if they would be willing to allow you to put on a ferret show as an extra attraction, though, doing it this way, you will not receive the showground entry fee. You will be forced to survive off ring entry and racing entry fees. The venue will also need toilet facilities and running water, for the animals will need fresh water, though the exhibitors should really bring this with them and many will indeed do so. You do not want to be paying for the actual venue, so highlight how the owners will benefit financially from allowing you to use their premises and you will undoubtedly get the place free of charge. Your outgoings must be kept to a minimum and getting the venue without cost is just one way of doing this. Having sorted out a suitable venue, it is now time to go out and get sponsors for the actual event.

The Sponsors

Local businesses are the best source of sponsorship and it will cost very little to purchase a small trophy. It is the cost of the rosettes that is the biggest problem, though this cost can be shared by more than one sponsor, and possibly several. There are also costs for advertising and printing costs for schedules to be met, so, unless you have plenty of spare cash, a few sponsors will be necessary in order to get you off the ground. From then on, the money raised from ring entry fees and showground entry fees should easily pay for the next show, if, indeed, you are up to putting on another venue! It is essential to be smartly dressed when approaching a potential sponsor and make it clear that they will receive plenty of publicity through the event. I always approach the sponsors personally, for you will probably find that you get nowhere when sending letters, though it is important to telephone first and make an appointment with the manager, or owner, of the business, prior to making a visit. You can include the name of their business within the pages of your schedules and possibly in any advertisements made.

Advertising

This can be done in the pages of magazines which deal with the subject of ferrets, through club newsletters and notices can be put up in local shops and supermarkets. You can also check with your local radio station, for these may have announcements for forthcoming events which are given out free of charge, and ask them to put you on their list.

Stewards and Judges

Stewards should be chosen from reliable people who are not going to let you down. Judges should be experienced ferret handlers who are familiar with the show scene, and your local ferret club, or welfare society, will help you find suitable judges. It is good practice to offer your judges their expenses, for fuel etc, and to feed and refresh them with drinks free of charge. Of course, if your show is very successful and you make a large profit, then you can pay your judges a modest fee for their valued services, although, especially if your show is put on in order to make money for charity, or a ferret welfare centre, they will probably offer their services free of charge.

Fees

The showground entry fee should be modest. As a guide £3 per car and around 50p or £1 for the ring entry fee. This is not too much to pay at all, for the costs of actually staging a show and racing can be quite considerable.

Insurance

It is advisable to get insurance for your event just in case of any mishaps. This should be quite reasonable and any insurance broker will be able to advise you in this regard. Of course, you could hold an event without insurance, but you run the risk of claims being made against you if anything does go wrong, however unlikely this may be.

Before you are tempted to put on a show of this kind, get yourself very familiar with the show scene first of all, so that you know just what is needed and what is expected by the exhibitors, for, be assured, if you get it wrong the first time, then it is very unlikely that those people will come again and they will more than likely tell their friends not to bother. On the other hand, if you put on a good venue and you make certain that you have decent judges who know exactly what to look for, then they will come again and they will tell their friends to join them. So go to shows regularly and get the feel for these events, and then you will undoubtedly put on a show that will be worthy of the ferret scene, a scene which continues to grow in popularity day by day, week by week, year by year.

Chapter Fifteen

Ferret Clubs and Welfare Societies

It is true to say that in recent years ferrets have enjoyed a massive rise in popularity and the numbers of those now keeping them continues to swell dramatically. This trend looks set to carry on well into the future. With rabbit hunters and those who do not wish to reduce the rat population with poisons which will kill them slowly over several hours, ferrets have always been very popular because of their useful qualities which make them an ideal hunting partner. The increased popularity comes from those people who perhaps do not have room for a dog and who do not want a cat because of their independence and aloofness, and this can only be a good thing, though this also has its down side.

During the 1970s, in the British Isles in particular, it became very fashionable to keep big cats, such as pumas or panthers, as pets and this was a trend that continued to grow until it was necessary for the government to bring in legislation which would prevent this from continuing, for these creatures need specialized care and, of course, once adulthood has been reached, are a potential danger to the public should the animal escape. Though many owners were responsible enough to give up their animals to organizations which would ensure they were properly cared for, others, to their shame, reputedly released their charges into the wild and now there have been several sightings of big cats throughout the British countryside. True, none of these have been confirmed, but the evidence that big cats are now colonizing parts of the British countryside is rather compelling and cannot be ignored. If this is indeed the case, then those big cats must be the offspring of animals released into the wild when it became illegal to keep them as pets. Just how much of a danger these cats are to the public in general has yet to be discovered!

Some people though, after having obtained one of these big cats and having discovered that it takes a lot of time, effort and specialized knowledge to rear and look after them successfully, get bored and fed up and, because they do not know what to do with them, have been guilty of releasing them into the wild – a very stupid and irresponsible thing to do. It is the same with other exotic pets. This industry has taken off massively since the late 1970s in particular and many end up bored with the creature they have chosen and release them to fend for themselves, or the neglected animal ends up escaping.

Unfortunately, it also happens with ferrets. Many people begin keeping them with great enthusiasm and much diligence, but soon get fed up with them. If it was a hamster, then mum or dad would probably make sure it was okay, or it could easily be passed on to someone else, but, because of the myth of ferrets biting everything they come across, many folks won't even look at them, let alone take care of them, so they are often just released out into the wild to fend for themselves and the poor creature will either starve to death (if a fitch has been used for hunting purposes, then it will have a good chance of survival as long as there are enough rabbits in the area), be killed by a dog, or possibly knocked over by a car, or, as often happens, will wander into a residential area and, usually after causing a bit of a panic in the neighbourhood, will be picked up by someone (whose hands are usually donned in the thickest gloves which can be found), and this is where ferret clubs and welfare societies come into their own, for many ferrets will be passed onto these places and they will then get the care they deserve.

Often though, these ferrets that have been purposely released into the wild will bite, so gloves are probably necessary when handling them, for they will undoubtedly have suffered neglect and quite possibly wrong handling, or, from being a kit, have been handled very sparingly (and then only with thick gloves on), so a biter is thus produced. As I have said before, biters are usually made, rather than born, just as well socialized ferrets are made, rather than born. In fact, just the night before I wrote these words, I was reading an article in *The Countryman's Weekly* about a ferret that someone had found wandering around and that eventually ended up in the care of the writer of the article. This was a very vicious ferret and savagely bit both the finder and the article writer when he went to fetch it. He states that he never wears gloves when handling ferrets, but admits to resorting to a pair for this little biter which he nicknamed 'Sid Vicious' – the name catching my attention because of my brother's fitch of the same name, also a biter. I have no doubt that the writer will be able to tame this fitch until it no longer deserves the name, for he is an experienced handler and will sort it out in no time at all. Of course, it could be that this particular ferret has been living wild for a good number of months and so has forgotten its earlier contact with humans and is only behaving as any wild animal would when feeling threatened, but I doubt that this is the reason for its savage nature. It is much more likely to have been neglected, even abused, by its previous owner who no doubt released it once he had got fed up with having it around, when the poor creature no longer amused him, for this is usually the reason why adult ferrets resort to biting. Also, there are still a few misguided fools within the working ferret world who continue, despite all that is written by expert ferret handlers who work their animals, to believe that a fitch must be wild and vicious in order to be of any use as a working animal – what utter rubbish. I have seen ferrets sleeping on their owner's lap of an evening, and then they were out hunting rats and rabbits the day after. So disregard totally any advice which states that a ferret must be of a savage nature to be of any use as a working animal. It could be that the ferret which was the subject of this article, was a working animal that killed its prey deep underground and, after getting fed up of waiting, was left to its own devices, though the writer goes on to state that no one came forward to claim the poor fitch, something which lends much weight to the theory that it was released after becoming unwanted – the fate of many ferrets, sadly.

Whatever the reason for a fitch being out in the wild – perhaps it has been lost whilst out hunting rabbits with its owner (many will curl up and go to sleep after they have killed a rabbit in its burrow and some owners who must be in the minority, will get fed up of waiting and will leave their ferret, though many now use locator collars which enable them to be found and quickly dug out). Perhaps it has escaped from a house, or a cage, or it has simply been released after becoming another old, unwanted fashion item, whatever the reason, ferret clubs and welfare societies are there to pick up the pieces, for these will take them in, even the bad biters, and will do their utmost to rehome them. Of course, ferrets that come to them which are bad biters, probably because of wrong, even abusive, handling, will not be passed on for a time, possibly quite a long time in some of the more extreme cases, for they will need to be properly socialized and cured of their bad habits, so those who run these clubs and societies are persons who have much experience of handling ferrets and know just how to treat a biter, in order to get it out of resorting to biting every time fingers appear in front of it.

As you can imagine, the cost of running a club or welfare centre which takes in these strays and unwanted animals, is very high and that is just one of the reasons for running shows and ferret racing events, for this is a good way of raising funds and we can do our bit by supporting these events whenever possible. Also, if we are in a position to do so, we can volunteer our services for there are endless chores to be carried out when large numbers of ferrets are kept. Cages need to be cleaned on a regular basis, fresh food and water must be provided daily and, of course, the occupants need to be handled and exercised on a regular basis too so that they remain healthy and well socialized. When they are kept in large numbers, as is often the case, the workload will be hard and very time consuming, so it is good to volunteer assistance if at all possible. And it is a good idea for these members of welfare societies and clubs to do all they can to encourage youngsters to get involved, allowing them to help with the cleaning, feeding and handling, except, that is, for the biters which sometimes turn up after a spell out in the wild. These will need to be handled by an experienced keeper of ferrets, until the animal is once again tamed and has become trustworthy.

A well socialized, tame fitch is a joy to handle. Just the day before I wrote these lines I was at the home of a friend of mine, getting a few photographs of his ferrets which are both pets for his children, and working ferrets, for he helps keep the rabbit population down on a hill farm where rabbits do much damage to the pastures where the farmer grazes his sheep. Whilst I was there, Gerry asked me to hold his ferrets while he clipped their claws which had grown a little too long. The albino, a ferret Gerry has had for quite some time now, just allowed the proceedings to go on through to completion without fuss, but the younger ferret, marked like a polecat and very similar to one I had many years ago, wriggled for a time and even gently put its teeth around the finger of its owner, just to let him know that it wasn't too happy with the situation, before settling down and resigning itself to its fate. The next time its claws are clipped, like the albino, it will make no fuss at all. It is good to know that your ferret will not resort to biting when it experiences something that it does not like. As I have said before, working ferrets should be just as tame

The author's wife, Glynis, saving her kisses for another!

as those kept as pets. It was a delight to photograph and handle these two fine examples of the fitch family. Welfare centres will have many such ferrets and youngsters, providing they are not too young, will be able to handle such animals without fear of being bitten.

Ferret welfare societies, clubs and rescue societies, will do their utmost to re-home the many strays and unwanted animals that end up on their premises, and some of these centres are very successful indeed, managing to find good homes for literally hundreds of these poor fitches every year, which says a lot about their popularity as pets, show, racing and working animals. However, not all are happy with the working side of a ferret's life and there may be an increasing number of people running rescue centres who will not allow a fitch to go to a working home. This is ludicrous, for a ferret can enjoy the life of being both a family pet and a working animal, and it can enjoy both comfort and the use of its natural instincts. Ferrets also provide a valuable pest control service to farmers. Those running rescue centres must remember this!

Details of ferret clubs and welfare societies are included in Appendix II.

Chapter Sixteen

The Ferret Ban

At the time of writing, ferrets, unbelievably, are banned in some parts of the United States such as California and the five boroughs of New York, though ferrets are kept legally throughout the rest of New York State. I find this state of affairs very hard to comprehend and, though I have attempted to find out the reason for a ban, I have been unsuccessful as this information is not available at the moment, although, of course, things may have changed by the time this book has been printed and is on sale, and I can only hope that this is indeed the case.

One can only guess at some of the reasons for a ban on ferret keeping being imposed and irresponsible ownership may possibly have been a contributing factor. Though ferrets can sometimes escape from a badly constructed cage, it is the trend for keeping ferrets around the home that is the main cause for a ferret escaping and getting out into the neighbourhood. A fitch may well get out into the streets and among people who are not generally familiar with ferrets, some even mistaking them for a rat, and this can cause no end of problems and tall tales can often grow of a savage animal on the loose! If a ferret got loose inside a large apartment block, then this could create even more problems, for many do genuinely fear ferrets, usually because of a wrong image which has been promoted in the media for many years, and an escapee approaching a human in order to obtain a meal, may well be misunderstood and the person will probably think they are being attacked. It is my guess that escaped ferrets and some of the problems which have arisen as a result, has something to do with the ban imposed in some districts of America. No doubt the unjust image of fitches is also a contributing factor, and possibly a major one at that!

Of course, not all escapee ferrets are the product of irresponsible animal husbandry. Some very responsible owners, myself included, have had ferrets escape on occasion (this has happened to me just once and was discussed earlier in the book), maybe because of a part of the cage working loose over a period of time, or a hole that was unseen somewhere in the house, or maybe a window that was left open and was within reach. As I have said before, I much prefer to see a ferret caged and its exercise periods closely supervised, for this cuts down the risk of escape.

Much more than irresponsible ownership, or genuine accidental escapees, is the idiot who simply abandons their pet in a park or even on the streets, after they have become fed up with looking after the poor unwanted fitch, once the novelty has worn off. Sadly, this does happen with a very small percentage of ferret keepers, but, unfortunately, it is always the innocent who suffer because of just a few bad eggs in the basket. These are just some of the reasons which may be behind a ferret ban and the action taken by some is way over the top to say the least. There are many bad car drivers out there, but society doesn't ban cars. There are many bad builders out there, but we do not ban the building trade. So why ban people from keeping an animal which is a delight to own and brings much pleasure and satisfaction to thousands upon thousands of people?

It may be that the fear of these animals spreading diseases is partly responsible but this argument does not stand up to careful scrutiny, for a ferret can be innoculated against most of the illnesses that could cause problems to mankind, such as rabies.

Of course, as with any unjust legislation, this ban is being fought and, if it affects you, you can play your part by writing letters to those who have the influence to do something about it and you can support efforts to raise money which will be needed to get this law overturned.

If you live in an area where ferrets are banned, then the decision as to whether or not you will keep them is entirely up to you, but, be warned, if you do decide to keep them, then there is always the risk of your pet being confiscated. These may then be passed on to a rescue centre, or your fitch may even end up being put down. So make sure that you are always discreet, for whilst there are no 'ferret police' out there looking for illegally kept ferrets, it is more than likely that they will follow up any complaints they receive and you will undoubtedly have a visit from the 'F.B.I.' ('Ferret Ban Investigators!!!').

Being responsible and discreet means practising good animal

husbandry. For instance, if you live in an apartment block, then filthy cages and litter trays, or large numbers of fitches with their scent glands intact, will cause quite a stink which may reach the homes of your neighbours and this won't do a lot for your cause, nor will it keep the fact hidden that you are keeping ferrets in spite of a ban. Walking your fitch in the local park, or down the street will surely lead to it being confiscated, so you will need to keep the exercise area confined to your home. If you have enemies, then make sure these do not find out about your ferrets, for they will almost certainly report you.

We can only hope that common sense will prevail in the end, and that this ridiculous ban is soon overturned and the legalization of ferret keeping replaces it. Ferrets are misunderstood for sure and the fact that the Department of Health is enforcing this ban would indicate that they have fears for public health where ferrets are kept. What nonsense goes on in the corridors of power, for ferrets are no more a risk to public health than a dog, a cat, or, indeed, a budgie.

There is a danger, of course, that where one ban exists, others may follow, so it is important that ferret owners do all they can to promote the true image of ferrets so that these bans do not increase throughout America and, indeed, into other countries. Ignorance is mankind's worst enemy and it is undoubtedly ignorance that has led to ferrets being banned in some districts of the USA and it is this ignorance which must be overturned if common sense is to prevail.

Banning something only makes it more popular and this measure just will not work, in fact, many will be very curious as to why ferrets are banned and will no doubt end up keeping them because of the ban. This has its dangers, for, once the curiosity has been satisfied, many may become fed up with their pet and end up neglecting it, or even releasing it to fend for itself, though most, I am sure, would be responsible enough not to do this.

It is far better to spend money on promoting responsible care of animals, not just ferrets, than to resort to the ridiculous and desperate measure of banning them. It would be very interesting to know just how much money is spent each year on enforcing this ferret ban. It undoubtedly costs the taxpayer quite a bit, money which would be far better spent on the promotion of good animal husbandry. Even introducing a licence, or a permit, would be far

Responsible care of the ferret should be promoted as seenwith this albino.

better than imposing a ban, for the revenue raised from a measure such as this, could be used to cover the costs of this kind of promotion and education. I am sure that most ferret owners would be happy to pay a modest fee each year for the privilege of keeping ferrets, rather than having a ban imposed on them.

Appendices

I A–Z of Ferret Names

A : Alice (a friend of mine owned a polecat-ferret, or a sable, which he gave this name. It was a strange ferret which wasn't at all friendly, though it didn't bite, it was just very distant and a real loner. It was also a very long and slim creature), Addy, Asher, Ash, Alder, Amy, Adder, Arnie, Alto, Adel, Alf, Alfie, Alfred, Avens, Alvin, Aniseed, Archie, Adle, Addie, Al, Acky, Asaph, Asa, Aster, Angelica, Arrow, Abby.

B : Ben (after my ferret which was huge, with a massive head and he was possibly the most gentle ferret I have ever owned and one of the most playful), Benny, Battle, Bleak, Bruin, Blen, Bedale (from a small market town in North Yorkshire, England), Bruce (taken from Robert the Bruce, the Scottish King of ancient times), Briar, Blethyn, Blythe, Bret, Brian, Billy, Brock, Badger (after Kenneth Grahame's hero in *Wind in the Willows*), Basil, Beech, Berry, Bilberry, Brook, Burdoch, Burnett, Balsam, Bally, Bellis, Blackberry, Bramble, Betony, Bluebell, Broom, Bryony, Bugle, Bulrush, Butcher, Bess, Buster, Biddy, Bink, Boss, Bracken, Bella, Beck, Buck, Briton, Brittle, Bell, Baltimore.

C : Cassy, Cyril, Clement, Croft, Crofter, Coniston (taken from the village and lake which are part of the English Lake District), Candy, Crest (as in the crest of a hill), Corrie, Conny, Clancy, Conor, Crete, Campion, Carrot, Celery, Chaff, Charlock, Charlie, Chervil, Chicory, Clarry, Clover, Comfrey, Cockle, Corn, Cowslip, Cranberry, Chad, Chesham.

D : Dandy, Dusty, Derwent (after Derwentwater in the Lake District), Dak, Davy (after the legendary Irish piper and Low whistle player, Davy Spillane), Doug, Devon (taken from the beautiful English county), Dusk, Das, Dell (taken from Healey Dell, a beautiful wooded valley high up in the Lancashire hills), Dot, Dotty, Debby, Den, Denny, Dibble, Dribble, Drizzle, Daffodil, Dusky, Daisy, Dandelion, Dewberry, Dock, Dodder, Dodd, Dill, Dilly, Diamond.

E : Ellie, Exmoor, Eskdale, Eddie, Ebby, Elsie, Elsa (taken from the lion's name in the popular book and film *Born Free*), Etty, Elderberry, Elder, Elodia, Elymus, Erica, Ernie, Eric.

F : Fern, Fen, Fly, Flicker, Floss, Franny, Fran, Foxy, Frisk, Frisky, Fuss, Fussy, Fred, Freddie, Franky, Fennel, Flag, Flaggen, Flax, Fig, Figoro, Fleming, Fan, Farmer, Forester, Flick.

G : Ginny, Gem, Glen, Gill, Gilly, Gypsy, Ginger, Garlic, Goose, Griff, Griffin, Grouse, Gravy, Gentian, Gen.

H : Heather, Henry, Honey, Hatty, Hem, Hazel, Hawthorn, Holly, Hollyhock, Hyssop, Horseradish, Hastle, Hawk, Heath, Hemlock, Honeysuckle, Hetty.

I : Itsy, Icey, Ice, Ike, Ivy, Iris, Iggy.

J : Jill, Jilly, Jasper, Jenny, Jake, Jet, Jed, Jimmy, Jiggy, Jerry, Jess, Jessy, Jick (my brother, Mick, was going out with a girl called June at the time I obtained my most favourite ferret, and they put their names together and called her Jick, J for June, ick, for Mick = Jick, a good way of making up your own names), Jock, Juddy, Jelly, Jester, Jack, Jackie, Jade.

K : Katey, Kim, Kate, Kerry (taken from the incredibly beautiful Irish county), Kelly, Kendal (after the Cumbriam market town famous for Kendal Mint Cake), Kes.

L : Letty, Lucy, Lucas, Lentle, Lindy, Laddie, Lill, Lilly, Ling, Lady, Lavender, Lime, Linden, Liquorice, Lettuce, Lucerne, Luzula, Lyle.

M : Mantle, Monty, Massey, Morris, Moss, Mick, Mischief (a good name for a ferret which will undoubtedly get up to plenty of this!), Meg, Molly, Murphy (a good Irish name), Misty, Maple, Maggie, Mags, Mandy, Miner (another good name, for ferrets love to explore dark passages underground), Mint, Mona, Moley, Mell, Midge, Mac, Major, Marshal, Millie, Marsh, Madder, Mellow, Marigold, Marjoram, Mistletoe, Mustard, Musk, Malva, Marram, May, Meadow, Melick, Melliot, Mercury, Mars, Molina, Myrtle.

N : Nipper, Nippy, Nip, Nickel, Nigel, Nidge, Nelly, Nettle, Nelson, Nightshade, Nuphar, Nessy (after the Loch Ness Monster of course).

O : Oak, Oz, Ozzy, Oscar, Ox, Otter (a relative of the ferret), Onion, Oregano, Orris, Oat, Odentis, Orchid, Orca (after the killer whale), Oyster, Oxlip, Oddette.

P : Pike, Printer, Pennine (after the long stretch of bleak hills which stretches throughout northern England), Penny, Pat, Paginini, Patrick, Pobble, Preston, Piper, Pep, Percy, Punch, Pickle, Prickle, Parsley, Parsnip, Pepper, Poppy, Pansy (not for a male ferret, please!), Pearl, Peppermint, Pet, Petty, Pimpernel, Pineapple, Pinky, Perky, Primrose, Prunella, Purple, Peter, Perriwinkle, Perry, Paddy, Pestle.

Q : Queenie, Quill, Queen, Quaker, Quake.

R : Rush, Rigg, Rags, Racer, Riff, Ruff, Rift, Renegade, Riss, Reece, Ricky, Rick, Rice, Red, Reel, Rebel, Rex, Rydal (a beautiful area near Grasmere in Cumbria), Rusty, Rocky, Roy, Rastus, Rufford, Ross, Ridge,

Ranger, Radish, Raspberry, Rhubarb, Rose, Rosey, Rosemary, Rue, Ralph, Rattle, Red, Rubus, Rye, Ricky.

S : Spout, Sparky, Sid (after Sid Vicious, though, of course, we do not want our fitch to live up to the second part of the name), Selwyn, Snap, Snapper, Snappy, Snip, Snatch, Sally, Spiffy, Stream, Sett (the home of a badger), Sherry, Shandy, Socks, Sam, Sammy, Storm, Stormer, Squeak, Stump, Sheena, Sedge, Smudge, Sting, Stinger, Scree, Slate, Saffron, Sage, Savory, Shallot, Shepherd, Silver, Sloe, Sorrel, Strawberry, Salvia, Syl, Samphire, Slender, Star, Snowdrop (a good name for an albino), Silas.

T : Timmy, Tess, Tipple, Tuppence, Tweed, Titch, Tickle, Tickler, Tarm, Tarn, Tatty, Teddy, Tracer, Topple, Tinker, Tan, Twist, Twister, Tig, Tigger, Trouble, Tip, Texas, Texan, Taffy, Tarquin, Tarka (taken from the name of the otter in Henry Wiliamson's timeless classic, *Tarka the Otter*), Tom, Tommy, Thomas, Trixie, Tina, Tangy, Tony, Tartar, Tyrant, Tricksie, Trick, Trinkett, Tack, Thorn, Tansy, Tarragon, Thistle, Thyme, Teasel, Thrift, Thrifty, Timothy, Traveller, Tutsan, Tway, Tween.

U : Ulpha (taken from a fell in the north of the English Lake District), Ulster, Uist (from north and south Uist in the outer Hebrides, where there are wild ferret colonies).

V : Vim, Vimto, Vixen (a female fox), Venture, Viper, Vick, Vicky, Violet, Verbena, Verb, Vernal, Vernon, Veronica, Vervain, Vetch, Viola.

W : Whip, Whippy, Whinny, Willow, Wren, Wist, Wasp, Wendy, Whisky, Woodruff, Woody, Warren (very apt, for ferrets have been used for ferreting rabbit warrens for centuries), Willy, Wily, Wilmot, William, Worry.

X : Xanadu, Xavier, Xeres, Xmas.

Y : Yorkie, Yorkshire, Yarrow, Yellow.

Z : Zena, Zak, Zealot, Zebra.

Naming a ferret can be quite difficult and one must rack one's brain to come up with a suitable choice for your fitch, one which suits it, so I hope the a–z of names will prove helpful in finding a name you are happy with. Of course, there are many other names which individuals will come up with and you may also think of many other names which are not listed here. I think naming a pet is very important, for the animal will be stuck with that name for the rest of its life, so it is necessary to get it right. Use the list as an aid, but try to think of other names to add to the list.

II   Ferret Clubs, Welface Societies and Suppliers

The following information is valid and correct at the time of writing and the websites and addresses given will help you to find just about everything you will need in order to keep your fitch properly cared for and thus in good, even vibrant, health. Of course, ferrets are so popular these days that most pet shops, even local smaller stores, will stock most of the things you will need, cages, food dishes, water bottles, bedding, litter trays etc. But there are many other suppliers and some of these have been listed here. At these places you will not only be able to obtain the things you need, but will also find many things to do with ferrets, often fun things, such as gift ideas. Also, much information can be accessed on the internet, though you must be careful that any information and advice given is reliable and coming from an experienced source, for literally anybody can set up a website and so not all the information on the internet is trustworthy, so exercise caution.

Ferret Associations of America

http://www.ferret.net is the site to visit. Add /groups/usa and the clubs and welfare societies of America will come on screen. By clicking onto these individual websites, you will find out much information about them and how to join if you wish to do so.

The American Ferret Association
PmB 255
626-C Admiral Dr
Annapolis, MD 21401
Phone 1-888-FERRET-1
Fax: 516-908-5215
E-Mail: afa@ferret.org

Ferret Organizations of America

WISCONSIN
F.R.O.L.I.C.
These are in western Wisconsin and they are encouraging people to start up other branches.

TEXAS
Ferret Lovers' Club

NORTH DAKOTA
There is another branch of F.R.O.L.I.C. in eastern North Dakota.

PENNSYLVANIA
Pennsylvania Ferret Rescue Association
These run an adoption scheme and their website will give you details of just how you can avail yourself of this service.
Three Rivers Ferret Club is situated in Pittsburgh

OREGON
Oregon Ferret Association
You will find much information on upcoming events on this site. These very active members try to put on one or two events per month.

RHODE ISLAND
Ferret Association of Rhode Island
These also run a ferret rescue service, as do many of the clubs, and you can find out about their fundraising events on this site.

NORTH CAROLINA
TriFL. Triangle of Ferret Lovers in Raleigh/Durham/Chapel Hill, North Carolina.
These have a good guide to local ferret suppliers.
Western North Carolina Ferret Associations. These are located in Granite Falls, North Carolina. These have a ferret shelter and run regular shows.

FLORIDA
South Florida Ferret Club and Rescue
These run educational seminars on ferret care and efforts of this kind must be commended. Responsible ferret care, something which has given prominence throughout this book, is essential if the good image of fitches is to prevail.

NEW YORK
Western New York and Finger Lakes Ferret Association
These will help with info about the ferret ban and there are many other benefits to being a member.

INDIANA
Circle City Ferret Club
Running an adoption service, this club will serve those in Indianapolis, Indiana and the midwest.

ILLINOIS
Greater Chicago Ferret Asscociation
These run a shelter and rescue service and are well known for their 'Off the Paw' newsletter.

MARYLAND
Baltimore Ferret Club
You will find on this site plenty of info and even how to get vet discounts. They will also tell you how to join, as will all of these websites.

MISSOURI
Springfield Ferret Association
No, this is not the home of The Simpsons, but it is the home of a very enthu-

siastic club which runs rescue and adoption services. They also have monthly meetings.

MINNESOTA
Another branch of F.R.O.L.I.C. in St Paul.

MASSACHUSETTS
Massachusett's Ferret Friends
These have info on the bill which made ferrets legal in this state and they have an events calendar.

CALIFORNIA
Ferrets Anonymous
This organization is working hard to get ferrets legalized in California. They include a list of ferret-friendly vets to be found in this state on their site.
Californians for Ferret Legalization
Another club which is fighting the legal system as regards ferrets and the ban on them.
California Domestic Ferret Association
These will provide regular updates on the legal battle raging in California.
Weasel Web and Golden State Ferret Society
Much info on the legal battle and upcoming meetings.

CONNECTICUT
Ferret Association of Connecticut, Inc
16 Sherbrooke Avenue
Hartford
CT 06106-3838
Phone: 860-247-1275
Email: agruden@ferret-fact.org
These run a rescue service too and you can find out much more on their website.
The Cirle City Ferret Club can be contacted on the following address:
P.O. Box 2272, Indianapolis, IN 46202 USA
Phone:317-767-1758.

Ferret Organizations of Britain

www.ferret.net/groups/britain

National Ferret Welfare Society
Everything you need to know about both pet ferrets and working ferrets can be found on this site. You can contact them on the internet at this site, or you can telephone: 0151-494-9059.
Email ccpl:Globalnet.co.uk

Black Dragon Nationwide Ferret Rescue
These will take any ferret that needs a home, making the letting loose into the wild totally unnecssary and unjustified.

London Ferret Owners
Mainly for people who keep ferrets indoors, with advice on how to provide an escape-free environment in your home.

Banbury Independent Ferret Welfare
Companions choice ferret food can be obtained from this website and the proceeds help raise money for ferret welfare.

Bedford Ferret Welfare Society
These hold regular events such as showing and racing.

Bolton Ferret Welfare Society
Sheila Crompton
38 Eastbank Street
Bolton
Lancashire
BL1 8LT
This welfare society promotes both working and pet ferrets and they will sell products which help raise money for ferret welfare.

West Cumbria Ferret Society
7 Copeland Avnue
Smithfield
Egremont
CA22 2QT

Cambridgeshire Ferret Welfare and Rescue Society
These have a busy calendar and you will find much information on their website.

Scottish Ferret Club
Find out about forthcoming events in Scotland on their website.

Ferret Organizations of Canada:

www.ferret.net/groups/canada

Greater Vancouver Ferret Association
These put on regular activites such as parties, ferret walks and educational projects.

Alberta Ferret Society
Another busy group, you can find out much about their activities and forthcoming events on this site.

Manitoba Ferret Association
These will help you find a ferret-sitter and can get you discount at local pet stores.

Toronto Internet Ferret Group
You can find info on vets, local suppliers of ferrets and accessories and a list of rescue centres.

Ferret Organizations of Australia and New Zealand

www.ferret.net/groups/aus-nz/

Ferret Paws
These are based in Victoria and they run a ferret shelter and hold monthly meetings.

New Zealand Ferret Protection and Welfare Society
You will find info on the society's constitution and on upcoming events on this site. They also have a list of lost and found ferrets.

Ferret Organizations of Europe

www.ferret.net/groups/europe

Belgian Ferret League
This site can be read in French, Dutch and English.

Danish Ferret Association
Ferrets are increasing greatly in popularity in many European countries such as Denmark.

Les amis du furet
Obviously a French club!

Frettchen Freunde an der Weser
A German club available only in the German language.

Associazione Italiana Furetti-Furettomania
An Italian club full of real enthusiasts.

164

Norsk Ilder Forening
A club in Norway.

Svenska tamillerföreningen
The Swedish Ferret Association is available on this site in Swedish and English.

Swiss Fancy Ferret Society
Though some is in English, most of this site is in German.

If you are not on the internet, then ask a friend or family member who is if you can use their facilities, or, alternatively, visit an internet cafe or a library. If you are still at school, or at college, then ask to use their computer.

Suppliers

www.theferretstore.com
You will find plenty of accessories on this site including some excellent cages.

Ferrets UK : www.ferretsuk.com
Telephone 01908 566366.

Arthur Carter
Landscove, Newton Abbot, S. Devon, TQ13 7NA. Telephone 01803 762127. Very good wooden carrying boxes can be obtained at this address.

Ferretworld Limited
Hotline: 01384-443533
Email. salesaferretworld.co.uk
Showroom 128 High Street, Kinver, Staffordshire, DY7 6HQ. 01384-87512.
Website: www.ferretworld.co.uk

Miscellaneous

International Ferret Welfare
Tel. 01922-627-029
Email. ferretwelofglobal.com

Star Ferrets
P.O. Box 1832, Springfield. VA 22151-0832.
Email. starferret@aeol.com

D.&R. Dunn
Ferret rescue and holiday boarding service covering Yorkshire and Licolnshire, England. Tel. 01964 614624 or 07773-754-269.

James Wellbeloved (ferret complete food)
Halfway House, Tintinhull, Somerset, BA22 8PA. Telephone 01935-410600. These supply Pet City, Kennelgate, Jollyes, Pets and Home and good local pet stores.

Index